PREFACE

These seventeen short science articles are derived from *Science Columns* that I wrote for the Vermont Standard newspaper over the past two years. There is no specific logic relating to the topics covered except that they are science related topics that I personally found fun, quirky, and of topical interest. I am very grateful to the editors of the Vermont Standard newspaper for granting me the rights to publish these articles independently, and I am grateful to my wife, Dr. Ruth Dlugi-Zamenhof, and to my friend of many years and published science author, Dr. Anthony Wolbarst, for encouraging me to write this book.

Happy bedtime reading!

Robert Zamenhof
South Woodstock, VT

TABLE OF CONTENTS

AIRPORT SCANNERS	3
GLOBAL POSITIONING SYSTEMS	9
100TH ANNIVERSARY OF THE THEORY OF RELATIVITY	14
HISTORY AND MATHEMATICS OF MAZES	21
IS RADIATION DANGEROUS?	30
RADIATION RISKS OF CT SCANS & THE BEATLES	36
RADIOMETRIC DATING	46
RECENT DEVELOPMENTS IN RADIATION THERAPY	50
COOL NEW APPLICATIONS OF WiFi TECHNOLOGY.	61
STEALTH MILITARY TECHNLOGY	67
UNDERWATER SUPERCAVITATION	75
DEFINITION OF TIME AND ATOMIC CLOCKS	79
IRAN NUCLEAR AGREEMENT	86
RED MERCURY HOAX	97
HORSESHOES & ABRAMS TANKS	102
DNA FINGERPRINTING	109
DARK MATTER	114
COLD FUSION—SOLVING WORLD ENERGY NEEDS?	116

AIRPORT SCANNERS

Since 9/11, the U.S. Department of Homeland Security has devoted large amounts of research to so-called 'threat-detection' at airports and other routes of entry into the U.S. If the only contraband objects were guns, the challenge would not be that great. Guns have traditionally been made of metal, which can be easily detected using specially adapted X-ray imaging devices. However, in 2013, a gun design was posted on the internet by *Defense Distributed*, an organization that posts digital designs of firearms that may be downloaded from the Internet and used to manufacture non-metalic firearms using 3D printing technology. The 'Liberator' is a one-shot .38 caliber pistol, named after similar pistols air-dropped by the Allies over France during its Nazi occupation in World War II. Unlike the original Liberator, however, the modern digitally built Liberator is almost entirely constructed of plastic with only the ammunition and firing-pin being made of a low-density metal. These properties make the Liberator very difficult to detect using traditional X-ray imaging.

As well as the challenge of detecting Liberator-type weapons, TSA staff are faced with the need to detect various kinds of plastic explosives, the most common of which are C4 and Semtex. Semtex was the explosive that brought down a Boeing 747 over Lockerbie Scotland in 1988. In addition to explosives, an important security goal is to detect illegally transported drugs.

The common feature between plastic firearms, plastic explosives, and illegal drugs is that they are all made of non-metallic, low-density plastic, or organic materials that can easily defy classical X-ray imaging technologies because they possess roughly equivalent density to human tissues. Recognizing this deficiency, two major developments in threat detection technology have occurred over the past decade: X-ray backscatter and T-wave imaging.

Whereas conventional, or 'transmission', X-ray imaging relies on sending an X-ray beam through an individual and detecting an 'X-ray shadow' image on the other side (like having a chest X-ray taken in a hospital), X-ray backscatter imaging uses a narrow raster scanning pencil-beam of X-rays that scatters backwards from an individual's body and produce a backscatter image of the body including any threat objects down to about a two inch depth. The main difference between these two kinds of X-ray imaging is that X-ray backscatter imaging is much more sensitive in detecting low-density materials such as plastic guns, explosives, and drugs. It can also detect denser metallic objects and can be implemented using very much lower and safer levels of radiation exposure than transmission X-ray imaging . The picture below shows a typical X-ray backscatter image of a traveler obtained at an airport.

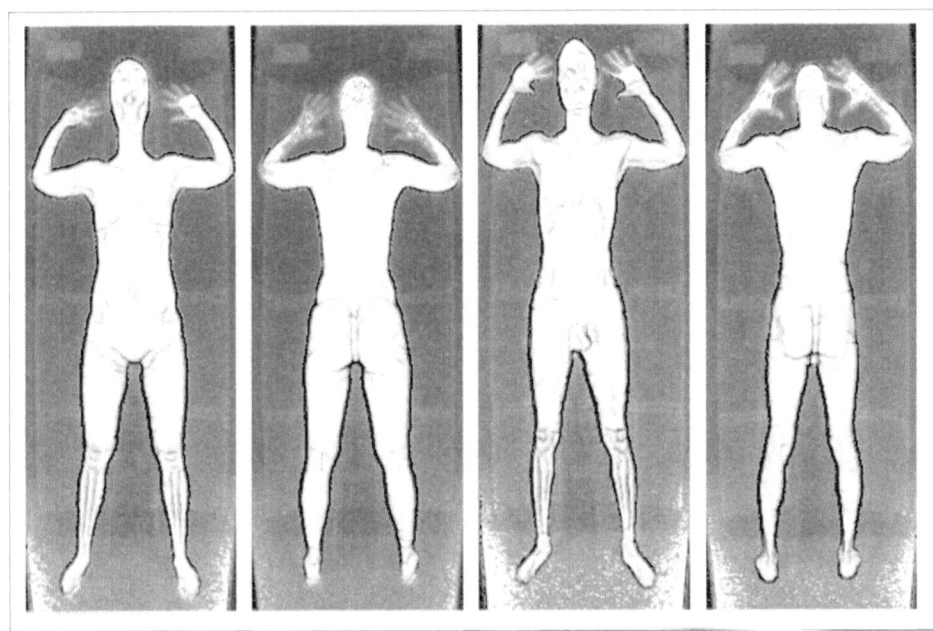

T-wave imaging stands for 'Terahertz-wave Imaging'— and is sometimes referred to as 'millimeter-wave Imaging' referring to the approximate wavelength of the radiation used. Instead of X-rays, 'active' T-wave imaging uses very low intensity beams of microwaves to create reflection images of threat objects hidden on a traveler's body, while 'passive' T-wave imaging uses the natural emission of microwaves from the traveler's body to create images of threat objects. The evident anatomical clarity of X-ray backscatter imaging is greatly superior to that of T-wave imaging, which is one reason why as a result of privacy issues T-wave imaging is now much more common in U.S. at airports and other threat-sensitive venues.

Added to this, there is the issue of radiation exposure to consider. X-ray backscatter imaging does involve a very minimal amount of radiation exposure; however, it has been shown that this exposure is no greater than the radiation exposure a traveler would receive from natural background radiation while waiting in line for 5 minutes to undergo such a scan. Put another way, the risk of death from cancer for an X-ray backscatter scan is comparable to the average risk of death traveling 100 feet in a motor vehicle. In contrast, T-wave imaging involves no radiation risk whatsoever; but, as mentioned, is markedly less effective for threat detection than X-ray backscatter imaging. The picture below shows a typical T-wave image of a traveler obtained at an airport.

With regard to privacy issues, both X-ray backscatter and T-wave scanners are able to provide images of the surface

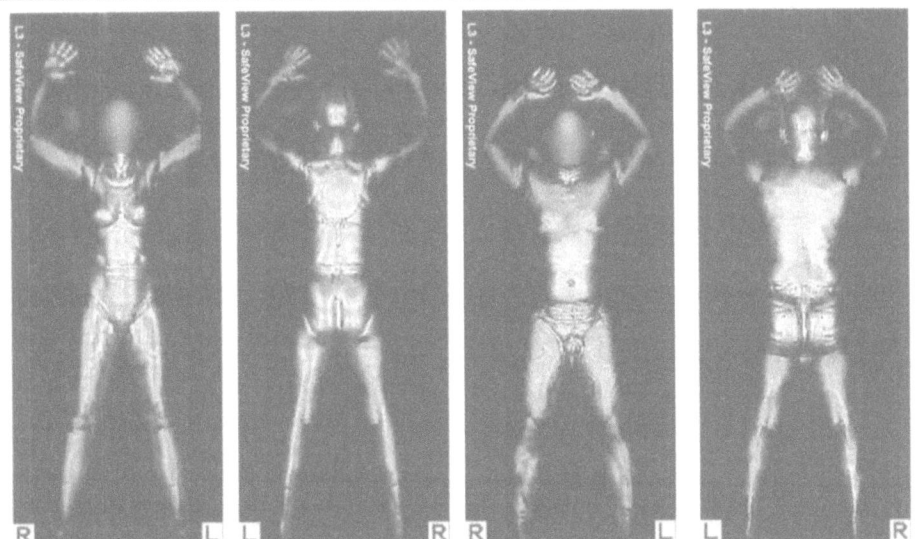

of a traveler's body; but, as already mentioned, the clarity of X-ray backscatter images is markedly better. TSA has implemented various solutions to depersonalize such images. For example, both a traveler's private parts as well as their facial features can be blurred using image manipulation software, while the TSA staff who examine these images reside in a separate room so that they see only the images and not the individuals being scanned. There are also more recent software applications that can automatically search for and flag threat objects in X-ray backscatter and T-wave images and only display the images to an operator if a threat object is initially identified.

There is no doubt that threat detection technologies have made a positive contribution to protection against terrorism, but unfortunately the public's misperception of the radiation risk involved with X-ray backscatter imaging and a needless concern with privacy issues has led to an almost complete substitution of X-ray backscatter imaging with less effective T-wave imaging technology at most U.S. airports and other threat sensitive venues.

In the interim, make sure you dispose of those small dangerous nail scissors before you travel by air. We know that an entire Boing 747 can be brought down by a passenger wielding a pair of nail scissors...

GLOBAL POSITIONING SYSTEMS

GPS systems have been a familiar consumer technology for the past couple of decades, but the great complexity of GPS is belied by its fundamentally simple principles. The U.S. GPS system, called NAVSTAR, uses 34 satellites—including 6 space-deployed spares. The 28 functional satellites are in precisely controlled orbits around the earth at a height of approximately 14,000 miles. In comparison, the international space station orbits the earth at approximately 270 miles. The picture below shows how the 34 NAVSTAR satellites orbit the earth.

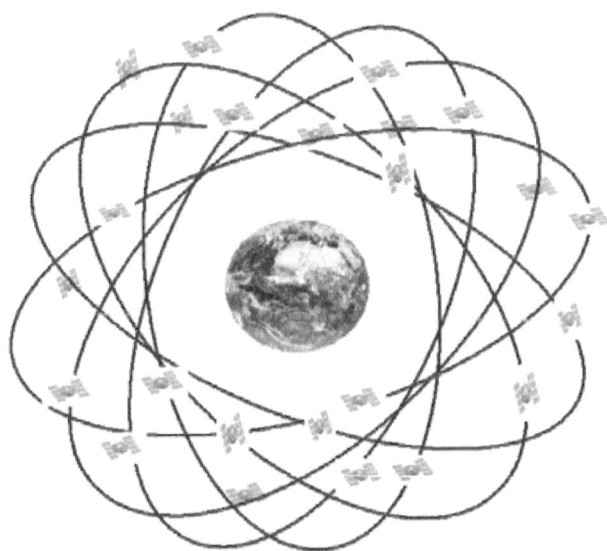

Each satellite makes a complete earth orbit every 12 hours. The satellites' orbits are continuously monitored by ground stations inside and outside the U.S. which very accurately determine the location of each satellite at specific points in time and can adjust their orbit if any corrections are needed. Each satellite periodically transmits a number of radio signals toward the earth which can be received by consumer, commercial, or military GPS units. These signals contain 'time-stamps', generated by phenomenally accurate atomic clocks on board each satellite, that document extremely precisely when each signal is sent out. With this information, a GPS receiving unit can calculate its distance from each satellite it 'sees'. With just a single satellite, the GPS unit must be located somewhere on the surface of an imaginary sphere of radius equal to the calculated distance of the GPS unit from the satellite. With four or more satellites, the intersection of the surfaces of four similar imaginary spheres determines the position of a GPS receiver anywhere on or above the earth's surface (i.e., in three dimensions). Still more satellites improve positioning accuracy through redundancy.

The picture below shows how GPS positioning is accomplished on the surface of the earth in only two dimensions (i.e., the surface of the earth is assumed to be smooth) using a minimum of three visible satellites. The earth is shown surrounded by three GPS satellites—1, 2, & 3—simultaneously 'seen' by a GPS receiver on earth represented by the small dot labeled 'position'. Using the time-stamped radio transmissions from each satellite based on its atomic clock, the GPS receiver calculates its distance from each of the three satellites, as represented by the radius of each of the three circles. The single point at which all three circles touch each other is shown by the dot and is the position of the GPS receiver on the earth's surface.

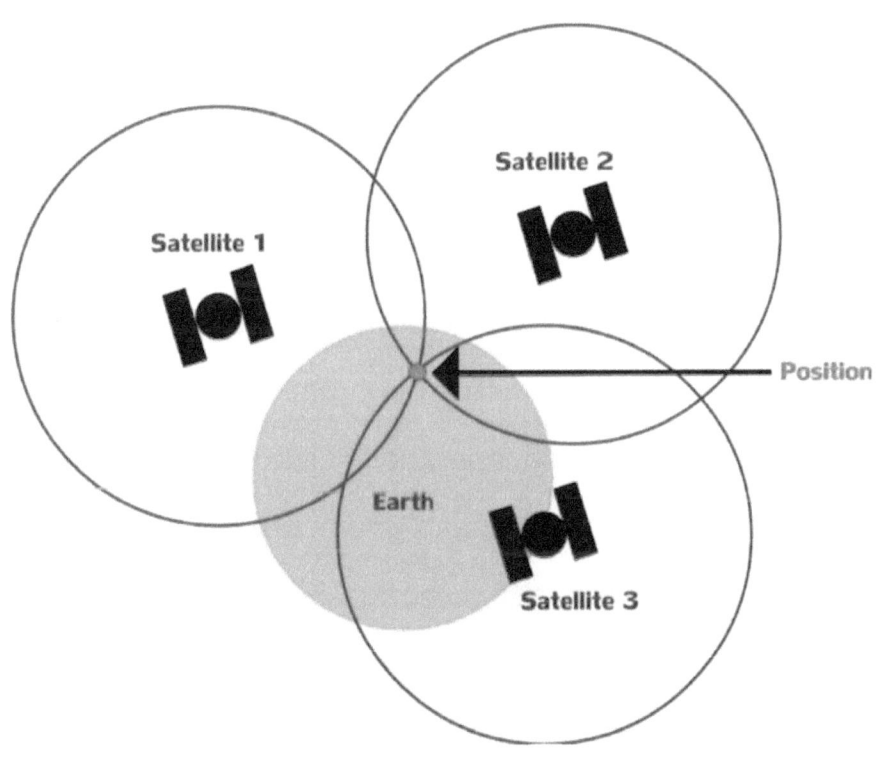

There are four principle sources of error in GPS positioning. Due to the satellites being near a huge mass—the earth—general relativity (see one of the other articles) causes the atomic clocks on board the satellites to very slightly lose time, which leads to some inaccuracies in the transmitted time-stamps. Because the satellites are moving very fast relative to the GPS receivers, special relativity causes a similar error in the transmitted time-stamps. When the transmitted radio signals pass through the earth's atmosphere, their paths are very slightly bent, which also interferes with accurate timing. And finally, the fact that the earth is not a perfect sphere, produces additional errors. Most of these sources of error can be adequately corrected for through theoretical adjustments.

In addition to the U.S. NAVSTAR GPS satellite system, which became operational in 1995, there are two additional global GPS satellite systems in operation: the GLONASS system in the Russian Federation, operational in 2011, and the GALILEO system in the European Union, operational in 2016. Japan, China, and India also recently developed their own GPS satellite systems, but these satellites are located in stationary earth orbit, and consequently can only provide positioning capability within those individual countries. Many GPS units, including consumer models, can use satellite data from both U.S. and Russian Federation GPS systems, which results in higher accuracy.

The accuracy of GPS systems generally depends on their design and intended application, but can be roughly summarized as follows. iPhone/iPad/stand-alone consumer GPS systems—are accurate to about 30 feet; professional mobile mapping and aircraft GPS systems--are accurate to about 6 - 12 feet; and military missile guidance GPS systems— are accurate to better than about half an inch. The construction cost of the U.S. and Russian Federation GPS systems was approximately $12-$14 billion (each), and the operating costs are a little under $1 billion a year.

GPS technology has revolutionized our ability to get places, target missiles, and keep tabs on the whereabouts of our dogs and cats. Micro-miniature GPS modules are under development and may soon enable us to track virtually anything that travels by mail, gets lost behind the refrigerator, or gets inadvertently put in the trash.

100TH ANNIVERSARY OF THE THEORY OF GENERAL RELATIVITY

Albert Einstein as a young boy.

In the early 1900s, Albert Einstein was at work in the patent office in Bern, Switzerland, and while looking out of the window at the building across the street and dreaming of his approaching *kaffe mit Schlag* (coffee with whipped cream) coffee break, he suddenly had a flash of insight: What would a person 'feel' if they jumped down from the roof of that building? Most of us would agree that a person jumping off a building would feel weightless; but what does 'weightless' actually mean? Weightlessness means that the person jumping, like an astronaut in space, does not feel a force pushing or pulling him in any specific direction. Conventional Newtonian mechanics and everyday observations dictate that the person jumping would accelerate downward at 32 feet/second/per second; while Newton's second law states that any acceleration requires the action of a force. So why would our person jumping *not* feel the force of gravity pulling him downwards?

This logical inconsistency led Einstein to speculate that gravity was not a 'simple' force—like the force of a strong wind or the feeling of a downward pull when carrying a heavy backpack--but was in fact a distortion in the 'fabric of space', which, like the side-rails of a bobsled run, would always steer a falling object toward the center of the earth, but not necessarily along a straight path. Einstein concluded that without this curvature of space there would be nothing to cause our jumping individual to accelerate—so space *had* to be curved!

What could cause such a curvature of space? Einstein

postulated that space would naturally curve near any large mass like our sun. A consequence of this would be that light rays heading through space to earth from distant objects, such as stars, and which would simplistically be expected to travel in straight lines would in fact curve inward as they grazed the surface of the sun due to the gravitational pull of the sun. Therefore, to an observer on earth, those stars would not appear to be in their locations as projected back along straight lines, but would appear to be very slightly displaced due to the curved trajectories of their light rays.

This new theoretical concept prompted Einstein to suggest to a colleague, British astrophysicist Sir Arthur Stanley Eddington, that they could test the 'curving light' theory of light by making measurements of the locations of stars during a full solar eclipse, when the moon totally intersperses itself between the sun and the earth.

Under full solar eclipse conditions, some bright stars located near the edge of the sun could be visible from earth because of the reduced glare from the sun as it is blocked by the moon. A few stars were chosen so that their light rays during the full solar eclipse would pass very close to the surface of the sun; and, if Einstein's theory was correct, the sun would bend their light beams inwards. As the eclipse receded and the earth moved so that the sun was no longer close enough to bend the light from those stars, the apparent location of those stars should slightly shift.

In 1919, with Einstein's encouragement, Eddington traveled to the island of Principe off the west coast of Africa to carry out this experiment. A full solar eclipse was to occur in Principe where observation conditions were expected to be favorable. The anticipated shifts in the locations of a few selected stars were calculated by Einstein and Eddington using Einstein's new theory. Using long film exposures from the start to the end of the solar eclipse, Eddington observed that instead of the selected stars appearing as dots on the exposed film, the photograph showed each of these stars as short horizontal lines as shown in the picture below. The large sphere is the moon and the short horizontal lines are the observed starlight shifts due to the curvature of space. Without the moon blocking the sun, the stars observed would be far too dim to be visible. The lengths of those short horizontal lines exactly matched the shifts Einstein and Eddington had calculated ahead of time. The success of this experiment made Einstein an overnight celebrity and the theory of general relativity ceased to be a theory and became an established scientific fact.

To illustrate how the fabric of space can be distorted due to general relativity, the rubber sheet analogy, shown in the next picture below, is often used.

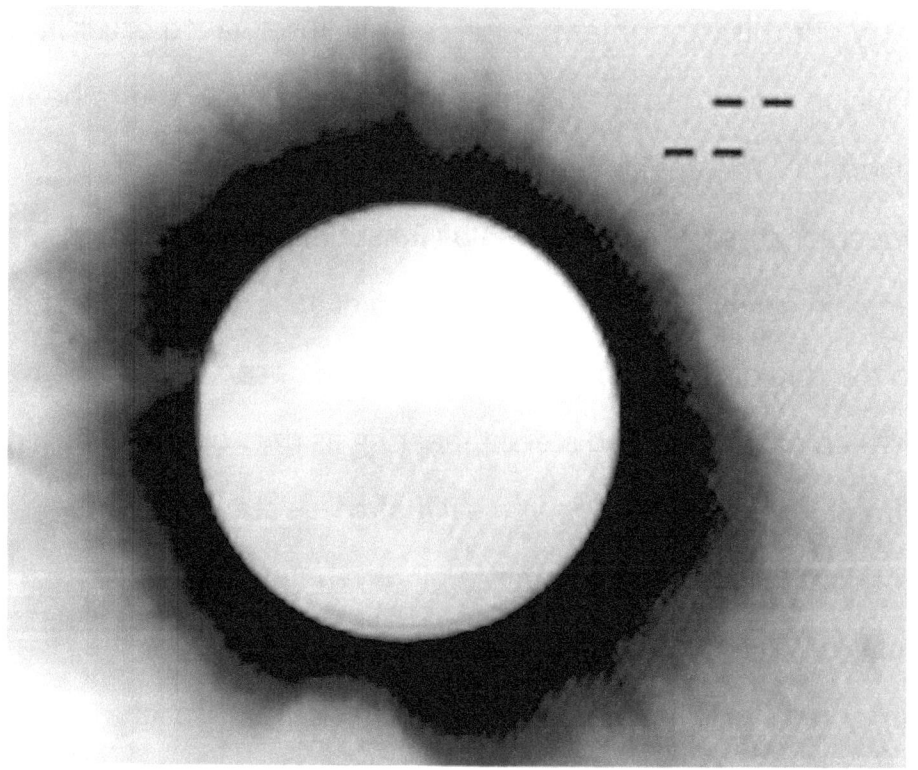

A stretched rubber sheet (like a trampoline) is marked with straight parallel vertical and horizontal grid lines that depict the undistorted fabric of space. A heavy ball is then placed at the undistorted center of the rubber sheet. Due to the weight of the ball the rubber sheet then becomes distorted in the area surrounding the ball as shown by the curved appearance of

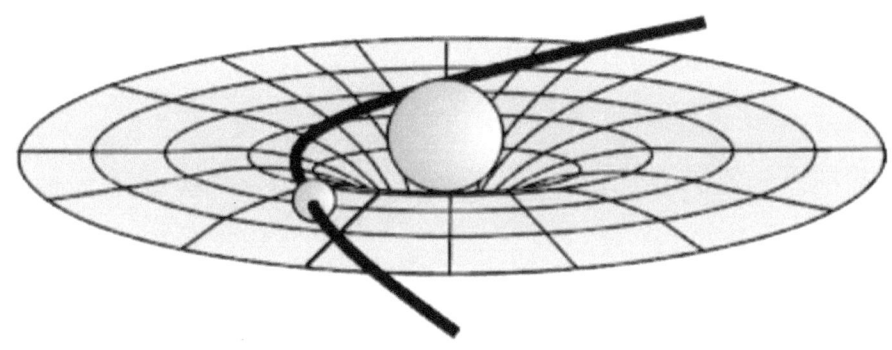

the horizontal and vertical lines. A small marble is placed on the rubber sheet near the ball and starts to roll along a trajectory that is 'bent' by the curvature of the lines on the rubber sheet. The marble rolls toward the ball roughly following the curved grid lines and illustrates what we now refer to as the 'pull of gravity' due to the curvature of space caused by any large mass.

There are many modern-day consequences of general relativity. Corrections have to be made to global positioning technology (GPS); the behavior of black holes can be explained; general relativity is the principle behind the operation of transformers and electric generators; and the expansion or contraction of the universe following the big-bang now makes sense (see the last article). And it was about 100 years ago that Einstein, looking out of his office window and looking forward to his *kaffe mit Schlag* had that brilliant inspiration that changed the scientific world.

HISTORY AND MATHEMATICS OF MAZES

The topic of mazes was an interesting one for the Church and for mathematicians in medieval times, and remains so today for families wanting an unusual activity to partake of on a Sunday afternoon.

The Ancient Greek historian Herodotus wrote in the 5th century BCE:

"This I have actually seen, a work beyond words. For if anyone put together the buildings of the Greeks and display of their labors they would seem lesser in both effort and expense to this labyrinth. Even the pyramids are beyond words and each was equal to many and mighty works of the Greeks. Yet the labyrinth surpasses even the pyramids!"

Herodotus was referring to the Egyptian Labyrinth, a colossal temple containing 3000 rooms on more than one level filled with hieroglyphs and paintings. Unfortunately, nothing recognizable remains of this ancient structure.

The Cretan Labyrinth was an extensive maze of corridors built by Daedalus for King Minos of Crete around the 15th century BCE. A medieval artist's depiction of the Cretan Labyrinth is depicted in the picture below, but again there is no remaining evidence of this amazing structure.

Before continuing, let me introduce some basic maze terminology. The terms *labyrinth* and *maze* are interchangeable. A *unicursal maze* is one that has no branching paths. There are no options for selecting the path one takes through such a maze. Most unicursal mazes were built as design foundations for formal gardens. A *multicursal maze*, on the other hand, has paths that branch and sometimes exhibit dead-ends along the way. The pathways ending in dead-ends are referred to as blind alleys. A *braid maze* is one with branches but with no dead ends. Multicursal mazes are the challenging kind, usually built in the form of gardens, hedges, or walls, with one or more entrances arranged around the perimeter and the end of the maze located at its center. For reasons that are probably more psychological than design intended, reversing one's tracks through a multicursal maze to return to the entrance point has often proved to be much more difficult! A *theta* is a maze composed of concentric circles. Finally, a *perfect maze* is one that has only one pathway solution—and for this reason is usually the most difficult to solve.

During medieval times mazes frequently appeared in churches and cathedrals. These were not meant to be physically negotiated but were often presented as friezes, inlays, or paintings on walls or floors ranging in size from about 5 ft - 50 ft. It is not entirely clear what purpose they served, but solving them—like a board-game--might have been a form of mandatory repentance for sinners.

The Bayeux Labyrinth from the 12th century is shown in the picture below and is typical of the genre of church mazes.

Mazes constructed of tall hedgerows became popular in Europe during the renaissance. Examples of such mazes are the Hampton Court maze in England, shown in the next picture below, followed by a picture of the Versailles maze in France.

Bayeux Labyrinth from the 12th century.

Simple mazes are also sometimes used in a laboratory setting to measure an animal's mental capabilities, such as its ability to learn from experience. For example, a rat introduced into a maze with a food reward at the end (that it can smell) will initially follow a semi-random path to try and reach the food. However, after repeated attempts, the rat will start to learn the architecture of the maze and eventually deduce the correct path to reach the food. The number of such attempts prior to successful negotiation of the maze becomes a measure of the rat's ability to learn. The effects of psychological stress, drugs, etc., can be studied in this way.

Hampton Court Maze, U.K.

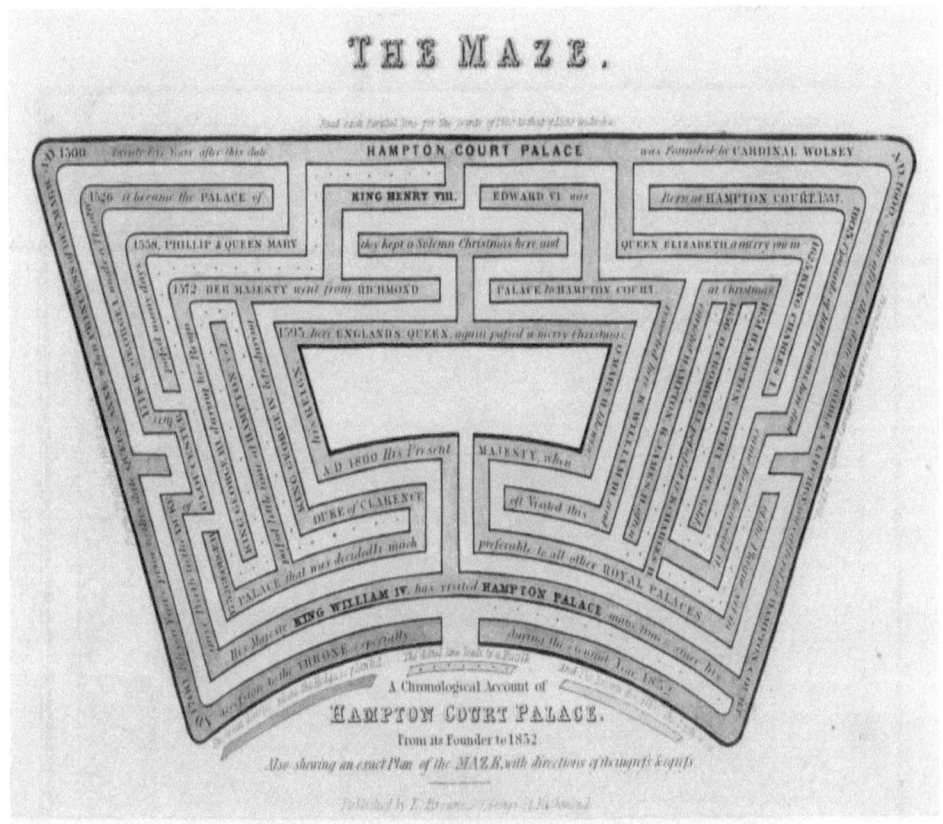

Charles Trémaux was a 19th century mathematician who devised a foolproof method for solving any maze. The picture below shows Trémaux's approach using a simple maze as an example. The small black circles shown are called nodes, where choices must be made of which way to turn. A 'new-node' is encountered along a path that has never been encountered before, while an old-node is encountered

Versailles Maze, France 1677.

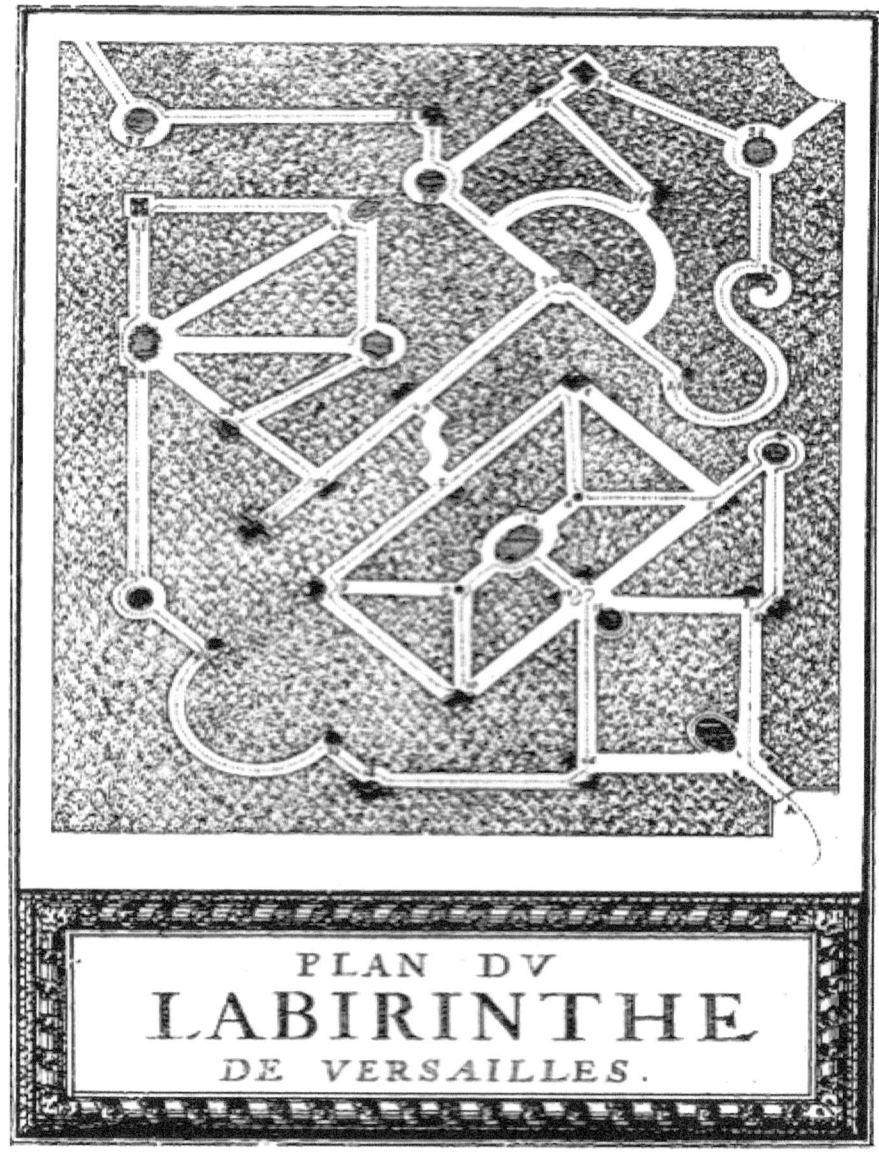

along a previously taken path. No path may be taken more than twice. When a new-node is reached, any path may be taken. When an old-node or a dead-end is reached, a new path must be taken (if one exists), or the old path arrived along must be retraced. These rules require you to either leave markings along your route or to take careful notes, but if Trémaux's rules are followed dutifully, one will eventually reach the end of the maze, although not necessarily by the shortest path.

Illustration of Trémaux's approach to solving any maze.

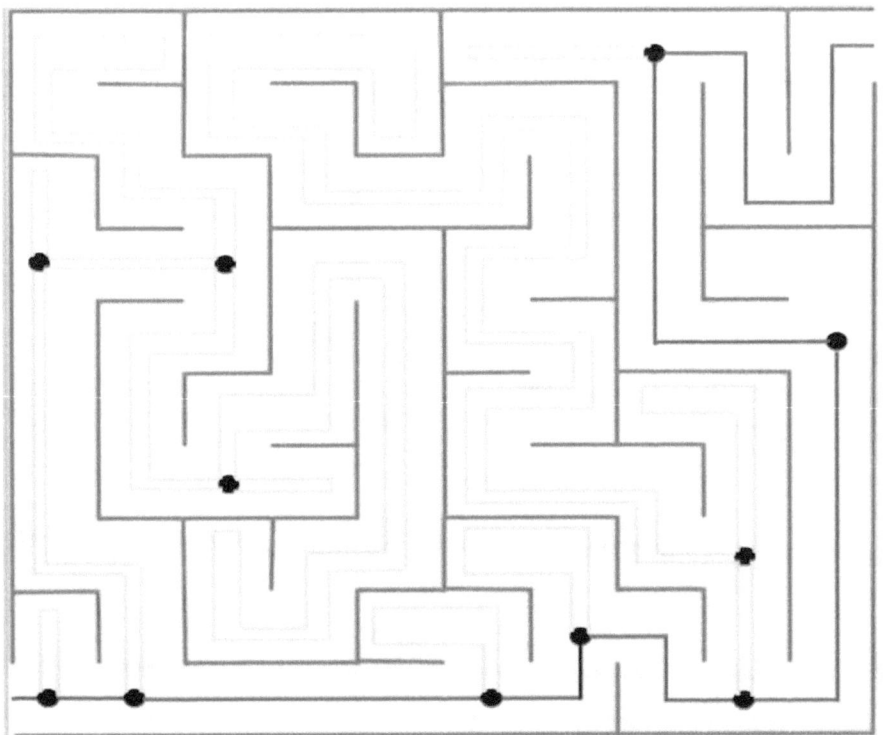

Trémaux's approach has been incorporated into some computer algorithms to solve maze problems available on the internet, although with current computing power most such apps use trial-and-error algorithms. A fairly complex multicursal maze is shown in the picture below. The solution, shown by the tracing, was computed using a free app called *Maze Solver*. The entrance and exit points are on the left and right.

So before you enter your next maze, make sure you have a sandwich, a large thermos of coffee, and a cell-phone; you may need them! Or, figure out your path ahead of time using a maze solving app.

IS RADIATION DANGEROUS?

About 10% of genetic mutations in humans are believed to be caused by 'background', or 'naturally occurring' radiation exposure. Naturally occurring radiation comes from cosmic rays from outer space (producing higher exposures at higher elevations), gamma-rays produced by various radioisotopes naturally present in the earth (dependent on specific geological characteristics), radioactive radon gas oozing out of the ground (produced by radioactive decay in the ground of the radioisotope radium-226), and radioisotopes that normally live inside our bodies (principally potassium-40).

Since evolutionary advance of the human race is driven by genetic mutations, naturally occurring radiation would not appear to be such a bad a thing. However, it is a two-edged sword: it helps drive evolution, but can potentially cause illnesses such as cancer and mental retardation in children.

Many of us are also exposed to man-made medical x-rays—both diagnostic and therapeutic. The average U.S. medical radiation exposure is numerically similar to the average U.S. naturally occurring radiation exposure. The picture below shows a naturally occurring radiation map of the

U.S. that shows natural radiation hot-spots in Nevada, California, Arizona, and New Mexico, where average radiation exposures are almost 10 times higher than they are in Florida. Scattered radiation hot-spots in Colorado and Wyoming are mainly the result of their high elevations, which increases the

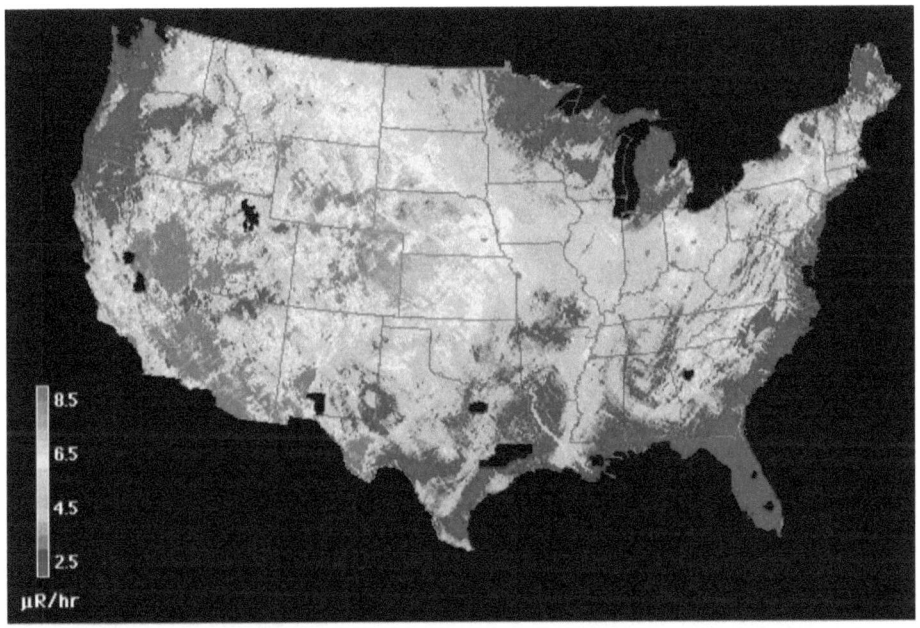

cosmic ray component of the naturally occurring radiation.

We know from countless studies that at certain levels radiation can cause cancer and that sometimes cancer can be fatal. Cancer, as far as we understand, is the result of genetic mis-programming caused by mutations in our DNA. Such harmful mutations can occur naturally, or they can be caused by external environmental factors such as chemicals or radiation.

The evidence we have linking radiation exposure to cancer deaths comes from historical population exposures to man-made radiation sources, such as the Hiroshima and Nagasaki atomic bombs, irradiation of the spine that many decades ago was a standard treatment in the U.K. for a congenital disease called ankylosing spondylitis, and exposure of female breasts in tuberculosis sanatoria in Massachusetts and Canada many decades ago, where a standard—although totally misguided--approach to the treatment of tuberculosis was to periodically deflate and re-inflate a patient's lungs under X-ray imaging guidance. Using all these data points, statisticians drew a straight line, originating at zero radiation exposure, and passing through the average of the scattered data points at much higher radiation levels representing the additional cancer deaths observed in the irradiated population in subsequent decades. Since there were no data points at the low radiation dose levels produced by naturally occurring or diagnostic medical radiation, this straight line was only a statistical estimate of what additional risk of cancer death might be expected at these much lower radiation levels.

Nevertheless, based on this straight-line model, which is

still used for radiation-related policy decisions by the U.S. government, the theoretical risk of cancer death for a U.S. citizen from one year's worth of average naturally occurring radiation exposure and an average annual amount of medical radiation is about 1 in 3,300 or 0.03%.

In comparison, the 'baseline' lifetime cancer-death risk among the human population in the U.S. (largely due to causes other than radiation exposure) is about one-in-5 or 20%, or about 0.4%/year. So if we believe the straight-line radiation risk model, every year one's likelihood of death from cancer would increase from about 0.4% to about 0.43% due to average natural and medical radiation exposure—not a tremendously shocking increase.

Another way of looking at this is to compare the 0.03% to the actuarial risks of partaking in other 'risky' activities. Such a comparison is shown in the table below, where the percentage-odds-per-year for death are listed for radiation exposure and for other risky activities. The statistical risk of death due to average natural radiation plus average medical radiation is marginally higher than that of – for example -- drowning in a swimming pool.

However, the straight-line model for estimating

radiation risks has inherent flaws. First, the data points that produce this line are so scattered that any predictions derived have very large statistical uncertainties—at the low radiation levels we have been talking about, these uncertainties could be as high as +/-100%. Second, compelling evidence has appeared in recent years to suggest that at the low radiation levels characteristic of naturally occurring and medical radiation, the incidence of cancer deaths is not only subject to enormous statistical uncertainties, but may in fact be 'inverted', i.e., *fewer* cancer deaths may occur in the presence of such low levels of radiation than in their absence.

The scientific phenomenon describing this counterintuitive effect of radiation is called 'radiation hormesis'. There are now many well documented experiments that have demonstrated the existence of radiation hormesis, and scientists are starting to believe in its statistical predictions in preference to the straight-line model. For hundreds of years, mainly in Europe, patients with various diseases have been intentionally exposed to low radiation levels by spending time in caves where the air contains low concentrations of the radioactive gas radon, and frequent improvements in health have been observed. If the radiation hormesis risk model were eventually proved to be valid for assessing health risks from radiation, this could have a tremendous influence on the future devices and techniques that are developed for medical diagnosis, on the choices that society makes for developing clean energy—for example, the reduced risks we would ascribe to nuclear power generation could have a fundamental impact on many sectors of our economy.

The table below [Calculated by the author] is based on 2017 actuarial statistics from the U.S. federal government showing cause of death vs. per-cent-odds-per-year for various 'risky' activities, compared to natural-plus-medical radiation exposure utilizing the straight-line model.

CAUSE OF DEATH	% ODDS PER YEAR
Accidental poisoning by and exposure to noxious substances	1.56
Drug poisoning	1.47
Opioids (including both legal and illegal)	1.04
All motor vehicle accidents	0.97
Car occupants	0.17
Motorcycle riders	0.12
Pedestrians	0.18
Average baseline cancer death-rate	0.40
Assault by firearm	0.35
Exposure to smoke, fire and flames	0.07
Fall on and from stairs and steps	0.06
Average Background and medical radiation	0.03
Drowning and submersion while in or falling into swimming pool	0.02
Fall on and from ladder or scaffolding	0.01
Air and space transport accidents	0.01
Firearms discharge (accidental)	0.01

RADIATION RISKS OF CT SCANS & THE BEATLES

Many years ago, a physicist colleague and I attended a conference at the University of Wisconsin in Madison. On the first morning, while we were taking our showers in the dormitory assigned to the conference attendees, the door burst open and 17 University of Wisconsin cheerleaders burst in. My colleague and I grabbed our towels and maintaining the highest level of decorum we could muster under the embarrassing circumstances, beat a hasty retreat. My colleague turned to me and whispered, "If you tell anybody about this I'll kill you!" Actually, my colleague—and friend-- was Professor Alan Cormack, subsequently the co-recipient with Dr. Godfrey Hounsfield from the U.K. of the Nobel Prize in Medicine awarded for the invention of computer-assisted tomography—or CT.

A few years after our encounter with the cheerleaders, the *Archives of Internal Medicine* journal carried a report that caused quite a stir in the media and the medical community. The report concluded that the 57 million CT scans done in the U.S. in 2007 could be expected to cause on a statistically predictive basis 14,500 future cancer deaths within the scanned patient population.

The statement that 14,500 future cancer deaths might result from 57 million CT scans seems like a very frightening thought, unless one recognizes that diagnostic uses of radiation aren't the only causes of cancer. There is also a so-called 'baseline fatal cancer rate', due to various factors such as environmental carcinogens, man-made carcinogens (food, drugs, chemicals, etc.), as well as, ostensibly, natural background radiation. This baseline fatal cancer rate in the U.S. and Europe is approximately 20%; i.e., 20% of the population eventually dies of cancer whether or not they have undergone CT scans.

Now the cancer deaths supposedly caused by CT scans, according to the Archives of Internal Medicine report, constituted only a 2% increase over this average baseline cancer death rate; i.e., increasing the risk from 20% to 22%.

Another way of looking at the predicted cancer death risk from CT scans is to compare it with actuarial risks of death from other common human activities. For example, the risk of cancer death from one abdominal, pelvic, or lung CT scan is statistically equivalent to the risk of death from smoking 220 cigarettes, drinking 360 bottles of wine, being exposed to air pollution by living in New York or Boston for 4 years, or traveling 40,000 miles by automobile. From that perspective, the risk of a CT scan—which is almost always associated with a corresponding improvement in a patient's medical diagnosis—doesn't appear quite as frightening or unreasonable.

However, there is an additional issue of uncertainty associated with predicting the risks of CT scans. An important issue of concern to radiologists and medical physicists—both experts on radiation effects associated with medical procedures—is the statistical model that is used to predict cancer deaths ostensibly caused by radiation doses at the levels used for medical diagnosis—including CT scans.

Most of the data linking radiation dose to cancer come from observations of the effects of the Hiroshima and Nagasaki nuclear detonations in 1945. Additional data came from studies of cancer in nuclear power plant workers in the U.K., patients exposed to x-ray fluoroscopy in U.S. and Canadian tuberculosis sanatoria, and x-ray treatments of patients with ankylosing spondylitis in the U.K. Typical radiation doses produced by current diagnostic techniques—including CT--are roughly 20-500 times lower than the lowest of the radiation dose levels in the historical data mentioned above. Extrapolating these historical data to the much lower dose levels used for CT requires an assumption of the shape of the dose-risk relationship.

A commonly used model that assumes a linear relationship between dose and cancer deaths is referred to as the 'linear-no-threshold-model' (LNT model). This model has been used for decades by U.S. federal agencies to quantify the risks of radiation dose and to set public policy relating to workplace safety and permissible diagnostic radiation dose levels on the basis of these predictions. The LNT model is a simple one: it assumes a straight-line relationship to exists between radiation dose and cancer deaths. Professor Carol Marcus of UCLA, an internationally respected authority on radiation risk assessment, has called the LNT model 'scientific baloney'; while Professor Ludwig Feinendegen of the University of Dusseldorf, Germany, an internationally recognized expert on radiation effects, has stated that 'the LNT model is unsupported by any convincing experimental evidence'.

The opinions of Professors Marcus and Feinendegen reflect a growing body of opinion that at the low dose levels encountered in diagnostic radiology, rather than potentially *causing* cancer, can actually *protect* the body against cancer.

The model that predicts this surprising conclusion is called the 'hormesis model of radiation effect'. Once the hormesis model of radiation effect becomes officially accepted, it would result in fundamental changes to our patterns of medical care, energy generation, and many other aspects of our everyday life and economy. Therefore, at the present time, we really cannot confidently say whether CT scans actually cause or reduce the incidence of cancer.

And finally, some interesting trivia. It has often been claimed that the first patient scanned by the prototype CT scanner built by Sir Godfrey Hounsfield--the co-recipient of the Nobel Prize for the invention of CT 50 years ago at the EMI Company's laboratories in the U.K, was a young musician who was suspected of having a brain tumor. His name was Ringo Starr. This new technology showed that Ringo did not have a brain tumor. John Lennon described Apple Records, who managed the Beatles in those early days, as a place where 'anybody with a good idea can get funding'. That may sounds like a questionable business plan, but we might have the Beatles and Apple Records to thank for providing significant funding for the early development of CT. This connection between the development of the CT scanner and the Beatles has recently been questioned--but it nevertheless makes a good story!

The first picture on the following page shows the first EMI CT head scanner, installed in 1971 at the Atkinson Morley Hospital in the U.K. The next picture shows the First CT scan of a human head obtained with an EMI Mark-I head scanner at Atkinson Morley Hospital. The image required a scanning time of 3 minutes and a further 5 minutes to reconstruct the image from the raw digital data. The image is displayed on a 80x80 pixel matrix. A large egg-shaped tumor is visible. Even this early 3D imaging technology was capable of providing superior cerebral tissue differentiation compared to a plain 2D skull X-ray shown below, which shows no brain tissue differentiation.

Plain 2D skull X-ray demonstrating no cerebral tissue differentiation.

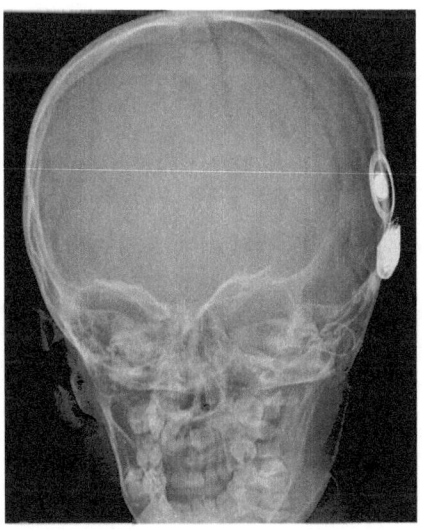

First EMI CT head scanner installed in the Atkinson Morley Hospital in the U.K. in 1971, with its inventor Sir Godfrey Hounsfield.

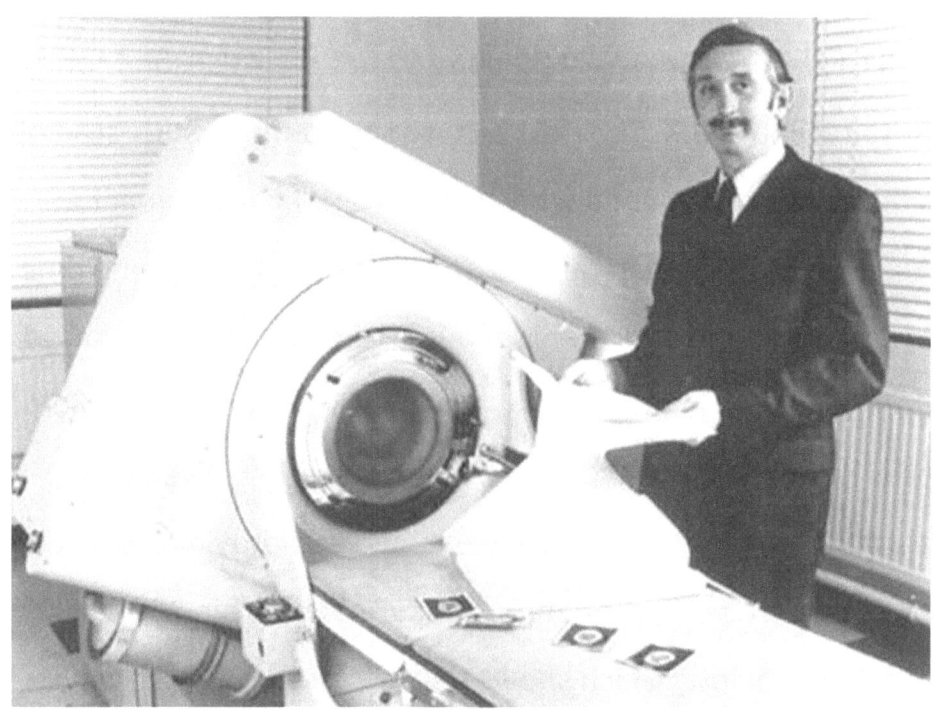

The picture on the next page demonstrates the historical progression in CT imaging sophistication. Shown is a modern CT scan of a patient's head obtained with a 10 second scan and 1 second reconstruction time. The image, from Cedars Sinai Hospital, has been specifically configured to emphasize the bony structures. A fracture of the jaw is clearly visible. In addition to medical imaging applications, CT has been used to examine mummified relics from ancient Egypt, density variation of sulfur filters in medical applications of nuclear

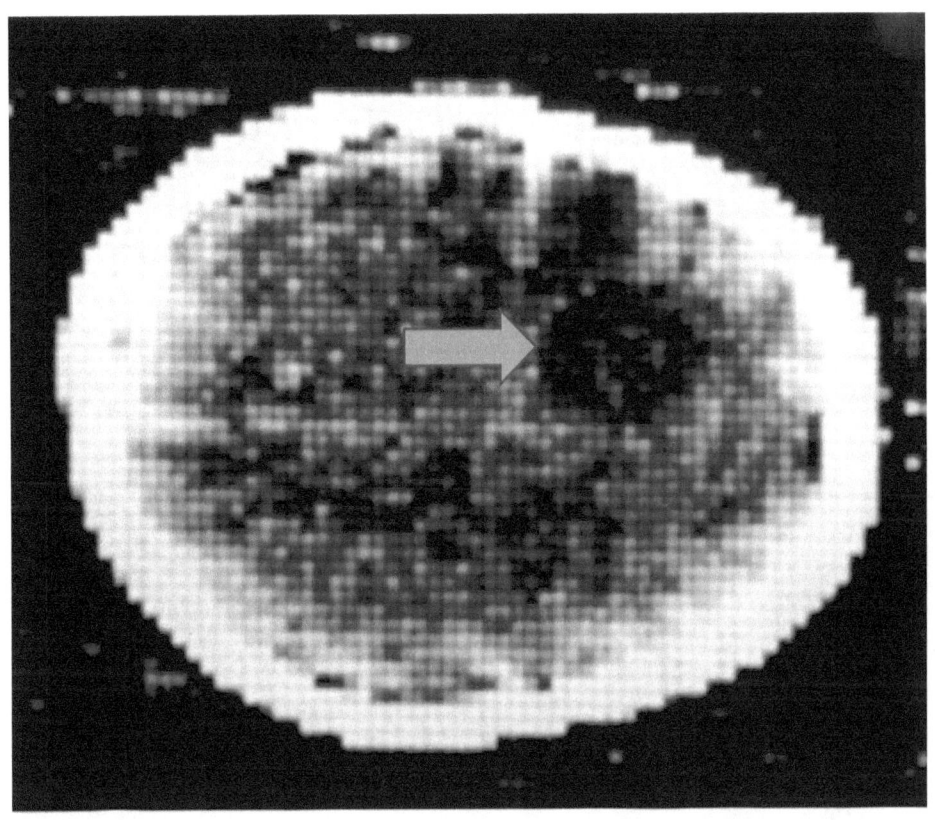

First CT scan of a human brain obtained in 1971 with the EMI Mark-I CT head scanner. An egg-shaped tumor is visible.

reactors, air bubbles in castings for aircraft engines, bone fractures in race horses, measurement of internal temperatures in patients, bone density in bone diseases, and much more. Two of the latest developments in CT technology have been a dedicated CT scanner for breast cancer screening and a portable dedicated mini-CT scanner for examining the legs of lame horses.

Modern era CT scan of the human head obtained at Cedars Sinai Hospital, CA with the software configured to emphasize bony structures.

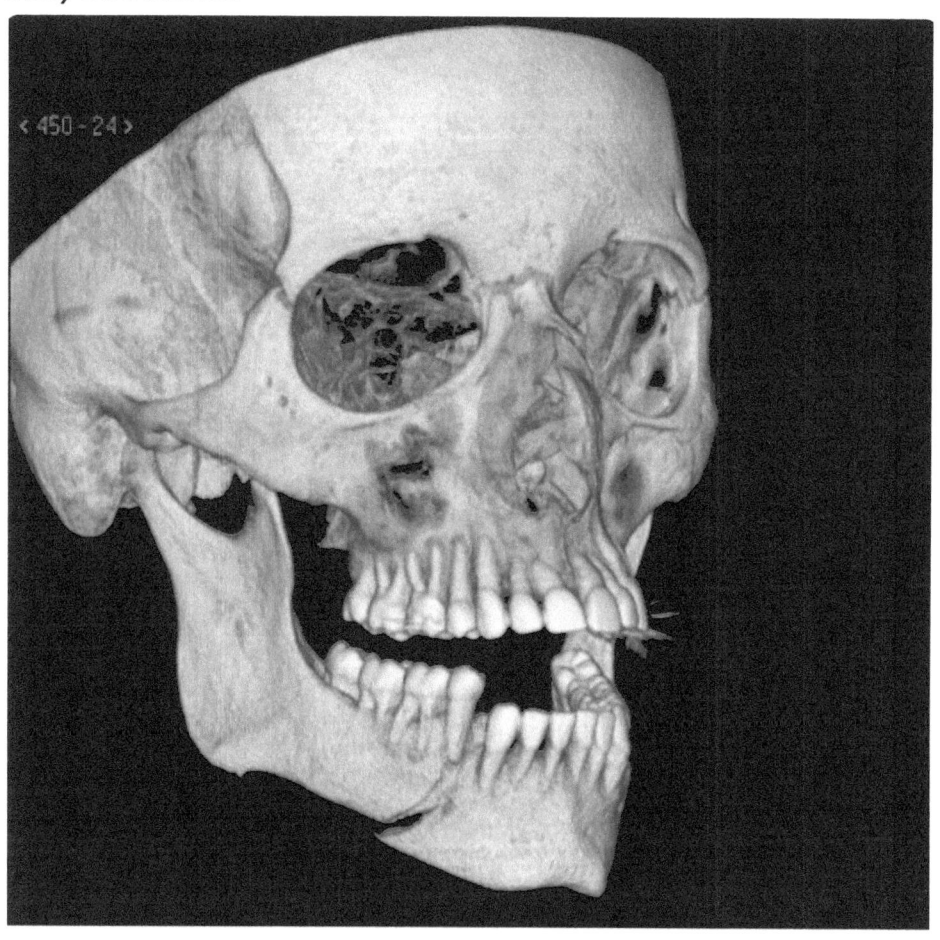

The final picture on the next page shows a CT scanner designed specifically for use on horses by the Prisma company. Up to now, to obtain a CT scan of a horse's head or legs, a horse had to be heavily sedated and a hoist used to guide the head or affected leg into the donut of a conventional CT scanner. The Prisma CT scanner, on the other hand, is 'open-plan' and can be used with the horse in a standing position without the need for heavy sedation.

Prisma dedicated horse CT scanner at the University of Pennsylvania School of Veterinary Medicine.

RADIOMETRIC DATING

Radiometric dating using radioisotopes was first proposed in 1907 and is now the primary method of determining the age of various objects. A subgroup of radiometric dating, 'carbon dating', is used to date a wide variety of plant and animal remains, and is a standard tool used by archaeologists and anthropologists.

The most common radioactive isotope found in the tissues of living plants and animals is carbon-14 (C^{14}). C^{14} is continually produced in our atmosphere by the interaction of neutrons (a major component of cosmic rays) with nitrogen. C^{14} then combines with oxygen to produce C^{14}-carbon-dioxide ($^{14}CO_2$), which eventually makes its way into living plants. $^{14}CO_2$ then enters the food chain and enters the bodies of living animals. $^{14}CO_2$ in living plants gradually decreases in concentration with time due to the radioactive decay of C^{14} (which has a half-life of 28,650 years) as well as excretion of C^{14} in the form of synthesized sugars; but $^{14}CO_2$ is also continually replenished by the intake of additional $^{14}CO_2$ through gas-exchange (plant 'breathing'). Eventually, a constant equilibrium concentration of C^{14} is reached in living plants. In parallel with this, a constant concentration is also reached of the most common *non*-radioactive isotope of carbon, C^{12}. But because C^{12} is not reduced by radioactive decay, the constant concentration reached by C^{12} is always *higher* than that of C^{14}. In animals, C^{14} and C^{12} are absorbed through the intake of plant food (which contain these two carbon isotopes) and excreted as CO_2 in breath. Therefore, the *ratio* of C^{14}/C^{12} in living plants and animals eventually reaches a constant equilibrium value.

However, when a plant or an animal dies, the intake of C^{12} and C^{14} abruptly ceases. While the C-12 present in the plant or animal at time of death thereafter remains at a fixed concentration, the C^{14} gradually decreases due to radioactive decay. Therefore, the *ratio* of C^{14}/C^{12} decreases with time after death of the plant or animal. By accurately measuring the C^{14}/C^{12} ratio, the time since death of a plant or animal can be calculated over a time scale of approximately 60,000 years.

Interestingly, carbon dating was more accurate prior to the 1950's than it is today, because above-ground testing of nuclear weapons by China, the U.S., and the Soviet Union altered the previously constant C^{14}/C^{12} ratio in our atmosphere. Although approximate corrections can be made, the ultimate accuracy of carbon dating became significantly poorer after the 1950's.

Other methods of radiometric dating rely on the absorption of elements other than carbon by living organisms or inanimate materials, and provide the ability to date animal and plant materials over much longer time scales than is possible with carbon dating, and are also used to date inorganic minerals present in animal bones and rocks.

Potassium–argon (K–Ar) dating is used most frequently in geochronology—the science of dating earth-rocks, moon-rocks, and geological events. The method is based on a very precise measurement of the conversion rate of the radioisotope K^{40} by radioactive decay into the stable gas Ar^{40}. Naturally occurring potassium contains trace quantities of K^{40}, and is a common element found in many materials such as micas, clays, and minerals. K^{40} converts to Ar^{40} gas, which is able to escape from these materials while they are in a molten or un-crystalized state, but not after they have solidified or recrystallized, at which time the Ar^{40} starts to accumulate and the ratio of Ar^{40}/K^{40} starts to increase. The time since a sample solidified or recrystallized can then be calculated by measuring the ratio of Ar^{40}/K^{40}. The extremely long half-life of K^{40} (1.26 billion years) enables this method to date materials from a few thousand years to a few *billion* years; in fact enabling the age of the earth itself to be measured as 4.54 billion years.

The picture below summarizes the principle of carbon- based radioactive dating. C^{14} is produced in the atmosphere by cosmic rays. The C^{14} in the atmosphere combines with oxygen to form C^{14}-CO_2 and C^{12}-CO_2. Plants absorb C^{14}-CO_2 & C^{12}-CO_2. Animals eat plants and absorb the C^{14}-CO_2 & C^{12}-CO_2 as well as breathing it out. When plants or animals die, intake of C^{14}-CO_2 and C^{12}-CO_2 and exhalation of C^{14}-CO_2 and C^{12}-CO_2 ceases. C^{14} decays and consequently the measured ratio of C^{14}/C^{12} goes down. The resulting C^{12}/C^{14} ratio is proportional to the age of the plant or animal at time of death.

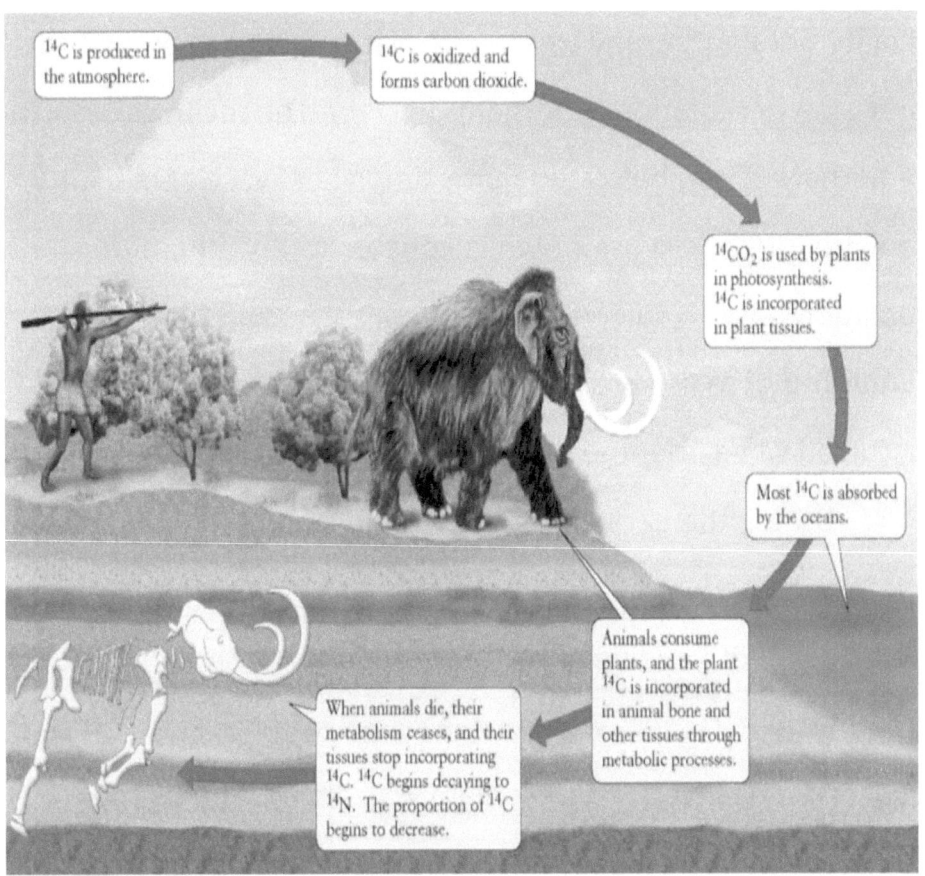

RECENT DEVELOPMENTS IN RADIATION THERAPY

Since their discovery in 1896 by Wilhelm Roentgen, x-rays and gamma rays have been the cornerstones of diagnostic radiology and radiation oncology. However, within the last two decades, new approaches to radiation oncology have been developed that go beyond just using X- or gamma-rays. Protons have now joined the radiation oncologist's armamentarium, while a much closer integration of imaging modalities such as CT, MRI, and PET-CT has improved the quality of radiotherapy treatment-planning and improved the precision of radiation treatment delivery. The technical goal in radiotherapy can be stated quite simply: delivery of maximum radiation dose to a tumor while concomitantly minimizing radiation dose to surrounding normal tissue and organs. A number of major breakthroughs in radiotherapy have occurred since Roentgen's time.

During the early decades of radiotherapy, the energy of the X-rays used was severely limited. This meant that X-ray beams did not have deep penetration, and consequently treating deep tumors would have raised skin doses to unacceptable levels. This problem was diminished by the introduction in the 1950s of the Van de Graaf electrostatic X-ray generator and the radioisotopes cesium-137 and cobalt-60. The higher energy of the gamma-rays produced by Cs-137 and Co-60 not only increased their depth of penetration, but shifted the maximum dose from the skin to a depth of a few millimeters below the skin surface, increasing skin tolerance to the doses delivered to deeper tumors.

In the 1970s a second breakthrough occurred with the introduction of medical linear accelerators. 'Linacs', as they are generally known, were again able to increase X-ray energy, further increasing the useful depth of treatment and shifting normal tissue doses to still greater depths below the skin surface.

During the early 1980s, three parallel breakthroughs occurred in radiotherapy: Faster computers and the introduction of CT scanners for treatment planning provided the ability to localize tumors and organs much more accurately, and in addition, the use of the anatomical data generated by CT scanners was used to more accurately calculate dose distributions within a patient's body. These developments resulted in safer treatments, superior tumor control, and the ability to subsequently evaluate the response of tumors following treatment.

In the early 1980s, the 'gamma-knife' was introduced, a device for treating small brain tumors, brain metastases (tumors from elsewhere in the body that spread to the brain), arteriovenous malformations, and some non-cancerous neurological lesions. In the gamma-knife, 201 small cobalt-60 sources emit narrow pencil beams of gamma-rays meeting at a single point in space. A patient's head is strategically moved so that this gamma-ray focal point is swept throughout the volume of a lesion being treated, producing a dose distribution that conforms very accurately to the geometric margins of the lesion.

The gamma-knife procedure requires a patient to have an aluminum frame mounted on his/her head, which involves four screws that penetrate through the scalp and into the skull. This procedure is done with local anesthesia and is not nearly as painful as it sounds.

The gamma-knife is designed to deliver single high dose treatments rather than the 20-40 treatment fractions comprising most radiotherapy procedures. In addition to being an effective treatment for intracranial tumors and metastases, the gamma-knife is frequently used to very effectively treat trigeminal neuralgia, a nerve disorder that can cause severe and disabling facial pain.

The increasing use of improved CT scanners and computers during the later 1990s opened up a new area in radiotherapy technology called 'image-guided radiotherapy' (IGRT). IGRT has made leaps and bounds since that time. For example, using a dedicated CT scanner mounted on the gantry of a linear accelerator, IGRT can correct for changes in organ position between treatment fractions, and can also follow the movement of tumors resulting from a patient's breathing *during* actual treatment. These factors both contribute to an increase in the ratio of dose delivered to tumor vs. dose delivered to normal surrounding tissues and organs thereby improving treatment outcome.

A different approach to radiotherapy, still only available in relatively few treatment centers, is called 'proton therapy'. Protons are nuclear particles that are accelerated to high energies by machines called 'cyclotrons' or 'synchrotrons'.

Unlike beams of X-rays that pass all the way through the body and deliver some unwanted dose to healthy tissues and organs beyond the depth of a targeted tumor, proton beams stop abruptly at designated depths. Consequently, protons can deliver more sharply contoured dose distributions, further increasing the tumor-to-normal-tissue dose advantage.

Protons are not advantageous for every type of cancer, but can be advantageous where tumors are in very close proximity to critically sensitive normal tissues such as, for example, tumors that are wrapped around the spinal cord or eye tumors adjacent to the optic nerve.

Another new form of radiotherapy that was introduced in the middle 1990s is called 'tomotherapy'. Tomotherapy still uses a linear accelerator as an X-ray radiation source, but unlike more conventional Linac treatment, tomotherapy operates more like a CT scanner. A couch moves a patient through a tunnel while multiple pencil-sized beams of X-rays are rapidly turned on and off in a carefully programmed sequence as the X-ray beams spiral around the patient. As with protons, many kinds of cancers can be more effectively treated using tomotherapy.

In the early 2000s, the 'Cyberknife' was introduced. This is a robotic radiotherapy system, programmed to aim a narrow pencil beam of X-rays into a patient's body from various directions, with the intersection of the pencil beams constituting the center of the point of treatment. Cyberknife treatments are especially useful for treating small tumors located within sensitive normal tissues. The Cyberknife is often integrated with a dedicated IGRT-enabled CT scanner so it can be programmed to follow the movement of a tumor during treatment. On a simple level, one may think of the Cyberknife as providing similar capabilities as the gamma-knife, but being also applicable to tumors outside the brain.

Let me end by describing what may truly seem like a science fiction radiotherapy technology that is currently being implemented in Finland in a large general hospital. The technology, called 'boron neutron capture therapy' (BNCT), was originally invented and developed in the U.S. The principle of BNCT is to initially infuse a patient having a tumor with a chemical compound containing the non-radioactive isotope boron-10. The chemical is specifically designed to concentrate in tumor relative to normal tissues and organs. Following the infusion of the boron-10--loaded chemical compound, the local area where a tumor has been identified is exposed to a beam of neutrons. Inside the cells, the neutrons are absorbed by the boron-10 isotope. Immediately following this, the boron-10 isotope disintegrates into two charged particles with ranges in tissue equal approximately to the dimensions of a cell (roughly

10 microns). Therefore, the additional dose delivered by the boron-10 to tumor cells that contain higher concentrations of boron-10 is largely localized within those tumor cells, effectively sparing surrounding normal cells.

Originally, neutron beams for BNCT applications were produced by modified nuclear reactors, such as the BNCT facility at MIT, but the clinical BNCT facility in Finland uses a particle accelerator together with an integrated CT scanner for treatment planning and delivery.

Clinical outcomes reported for earlier studies done by the Finish team were impressive for a number of different cancers including refractory head and neck cancers that had recurred following conventional treatment. The new accelerator based facility in Finland bypasses many of the regulatory hurdles experienced by reactor based BNCT facilities. One important benefit of BNCT is that isolated tumor cells surrounding the periphery of a tumor are treated by this form of therapy even if they cannot be directly imaged.

The pictures below show the technologies that have been described.

Elekta Gamma Knife patient treatment facility.

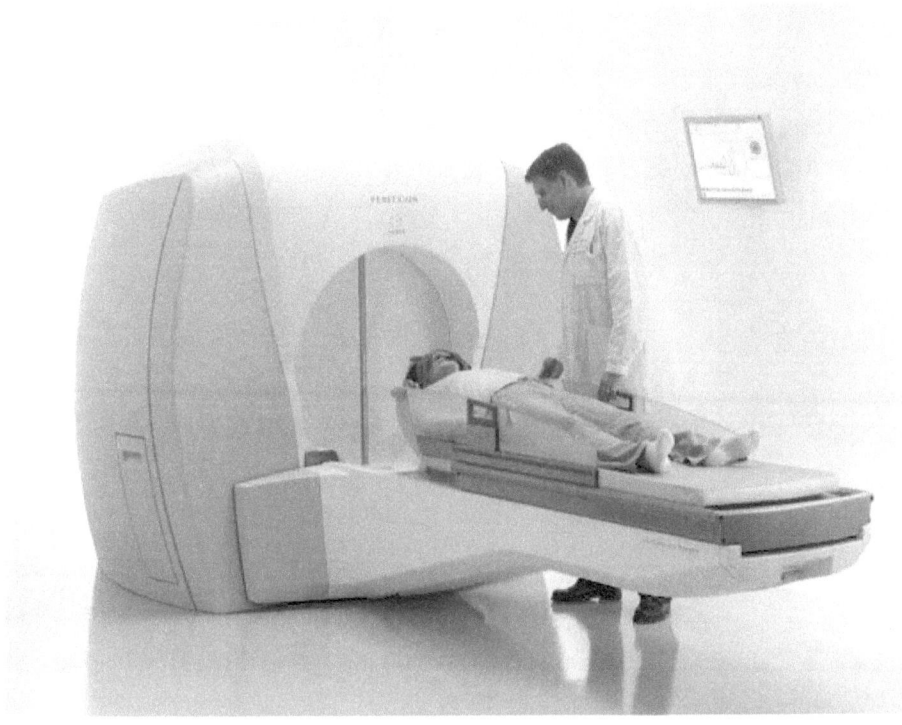

With the aforementioned advances in radiation therapy technology, the control of localized cancers has become extremely effective. To further reduce cancer mortality one needs the ability to target and treat 'invisible' cancer involvement, such as the diffusion of cancer cells to significant distances beyond visible tumor boundaries. The technology of Positron Emission Tomography coupled with CT is called 'PET-CT'. PET-CT – enhanced radiotherapy is a good example of the integration of nuclear medicine, X-ray imaging, and radiotherapy to optimize the success of cancer therapy.

Emory University proton treatment facility.

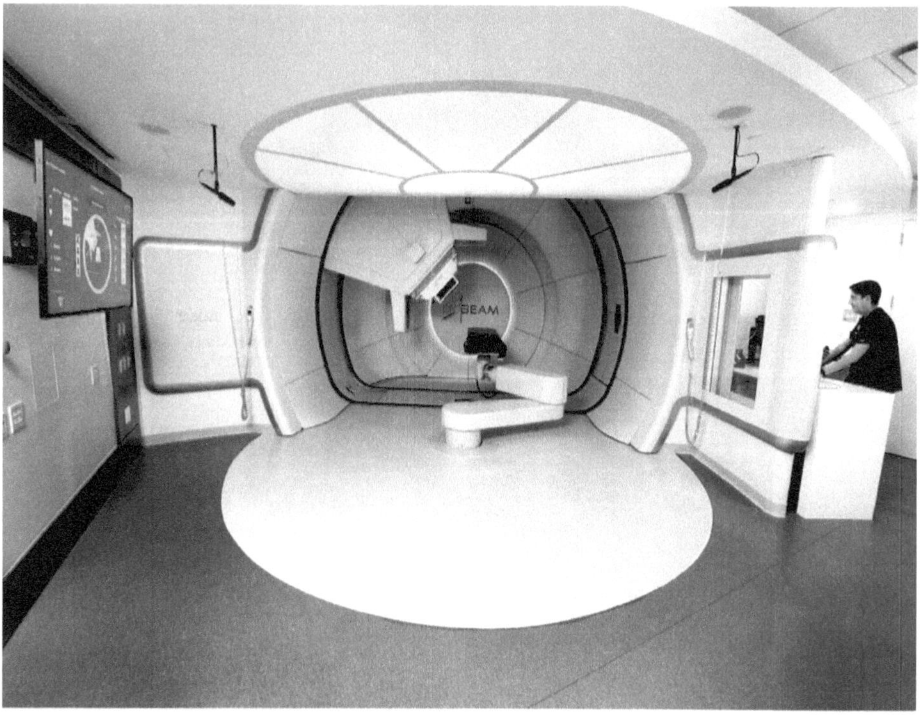

Linear Accelerator patient treatment facility.

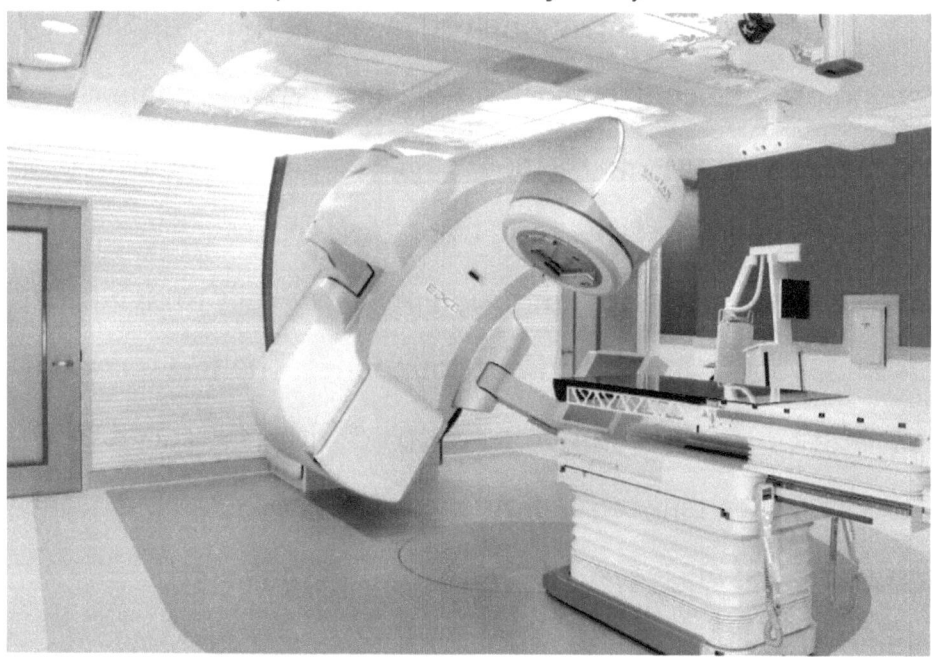

Cyberknife robotic treatment facility.

Tomotherapy patient treatment facility.

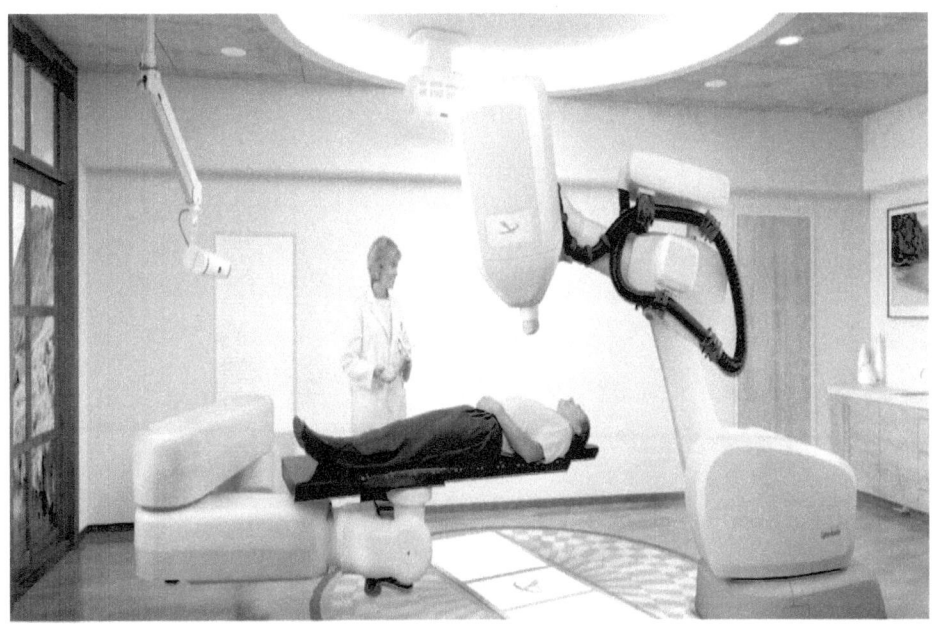

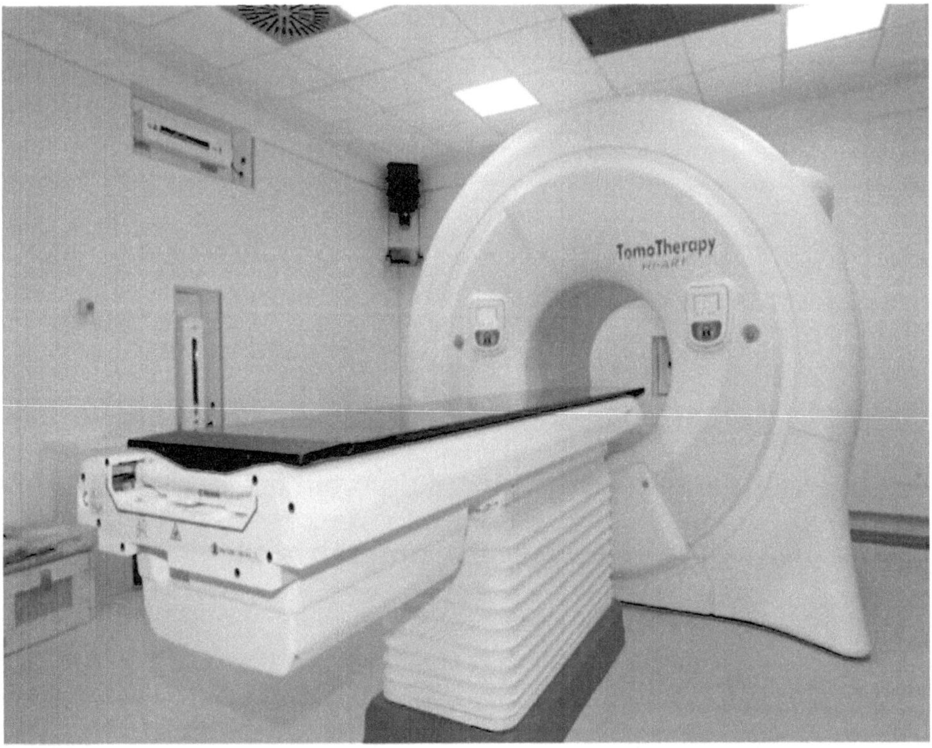

Experimental Neutron Capture Therapy patient treatment at the MIT Nuclear Research Reactor. Neutron collimator is visible in the background. Patient's head is held firmly in position during the 30 minute treatment period using a polyethylene net.
[Illustration provided by the author]

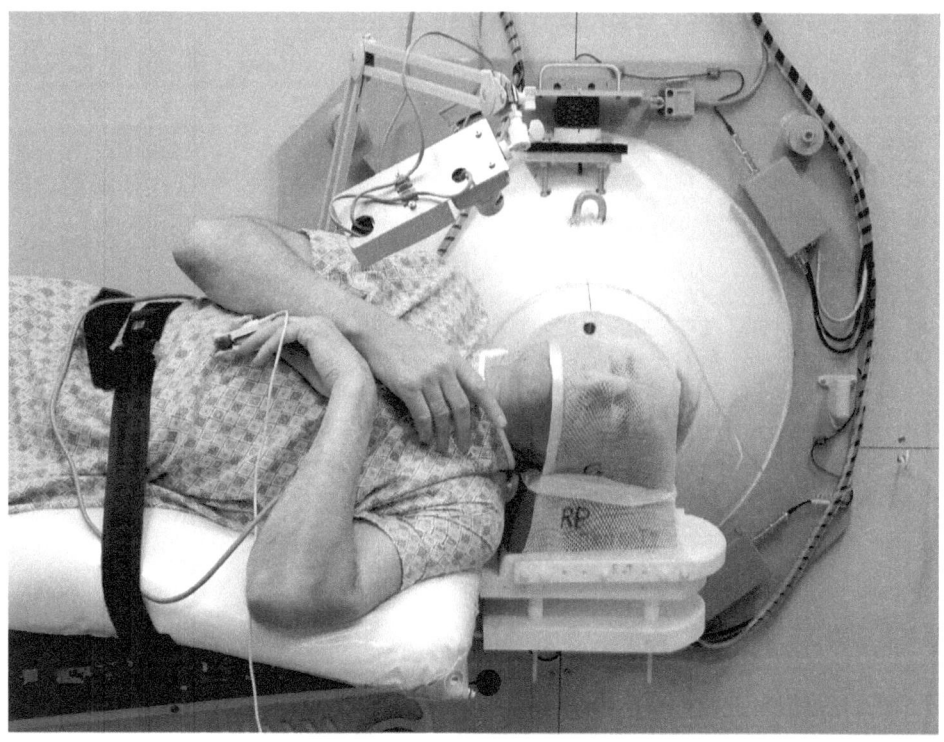

This facility, jointly operated by MIT and Harvard, was financially supported by the U.S. Department of Energy. Twenty-four patients with inoperable brain tumors were treated before the investigational program ended.

COOL NEW APPLICATIONS OF WiFi TECHNOLOGY

WiFi ('Wireless-Fidelity') replaces old-fashioned hard-wired 'Ethernet' connections between your computer and the internet access point in your home or business. For *outgoing* information, your computer's wireless module converts data into a radio signal, sends it to your computer's antenna, which transmits it to a 'wireless router'. The output side of the router is hard-wired to either a dish, telephone, or cable-based internet access point. For *incoming* data, the reverse sequence of events occurs. In the late 1990's, hard-wired computer internet connections morphed into wireless 'WiFi' connections, referred to as 'Wireless-Fidelity', based on Australian patents from the 1990's.

Most WiFi networks operate at a radio-frequency of either 2.4 gHz or 5.8 gHz ('g' means billion, 'Hz' means cycles/sec; so 5.8 gHz means 5.8 billion cycles/sec). This frequency is considerably higher than used for cell phones or TV transmissions. Higher frequencies provide more transmission 'bandwidth'—i.e., the capability of transmitting more data-per-second (expressed as 'bits/s'). However, the higher the WiFi frequency, the more difficulty signals have penetrating barriers such as doors, walls, etc., so the advantage of the lower 2.4 gHz WiFi signals is that they can penetrate more effectively to more locations within a building. But, by the same token, they are more vulnerable to hacking—it's a two-edged sword.

Three important parameters internet service providers advertise are *download speed*, *upload speed*, and *download data-cap*. If you use a WiFi connection to stream movies, you are only concerned with download speed. Service providers such as Netflix have guidelines for minimum necessary download speeds for movie streaming: 1.5 megabits/s for the lowest video quality, 3 megabits/s for DVD-equivalent video quality, and 5 megabits/s for high-definition video quality. But problems arise when multiple users are logged into a single WiFi connection, since the available connection speed for each user then shrinks proportionally. Such speed-sharing issues once again become more serious with public WiFi networks, where hundreds or thousands of users may be simultaneously connected and often doing a disproportionate amount of movie streaming.

Upload speed can become an issue if you are interacting with commercial websites, sending large email attachments, using online computer programs, or if you use the cloud as your computer backup. In addition, in many locations, download data-cap is a significant issue. One of the local internet service providers offers a pretty attractive transmission speed of 50 megabits/s, but caps downloads to 50 gigabytes of data each month. Depending on desired image quality, streaming a 90-minute movie requires about 1-5 gigabytes of data. With a download data-cap of 50 gigabytes/month, you are limited to about 16 movies/month in DVD-definition—assuming you don't download anything else. And that doesn't get you very far if you have to share your WiFi connection with your Brady Bunch doing their homework.

New applications of WiFi technology are continually under development. A relatively recent one is *in-flight satellite-based WiFi*. This uses the so-called 'Ku-Band' service currently offered by Panasonic. Unlike competing in-flight WiFi services, it provides WiFi connectivity to air travelers over land or ocean. Ku-Band antennas are housed on top of aircraft and are directed towards Ku-Band satellites linked to ground-based transmission stations. Although Wi-Fi speeds of up to 30-40 megabits/s are theoretically possible, actual connection speeds are substantially slower, depending on the number of aircraft that are simultaneously sharing the Ku-Band network.

Public WiFi signals are electronically 'stamped' with their source locations, so newer GPS systems are starting to use WiFi signals to further enhance their positioning accuracy, especially in locations partially shielded from GPS satellite signals.

To further secure WiFi connections from unauthorized access, specifically in the business sector, a development called 'LiFi' uses light instead of radio signals. LiFi coding is done with LED lights that would normally be used for room illumination.

The advantage is that LiFi-coded WiFi signals within an office space rarely penetrate to the outside, or even to neighboring office spaces, so the level of access security is greatly enhanced.

Airport screening for weapons, explosives, and other prohibited articles is also at the point of using WiFi. Since WiFi services are ubiquitously present, this new technology proposes using WiFi radio waves to produce images of certain illegal items carried by travelers (similarly to how medical X-rays produce images of body-parts). This application of WiFi promises to be at least as accurate as many of the threat detection technologies based on 'T-waves' currently used at most U.S. airports.

Researchers at North Carolina State University have developed miniaturized WiFi 'backpacks' that can strap onto the backs of cockroaches. These cockroaches, illustrated in the picture below, can be introduced into very small spaces from which they can transmit various data signals. For example, following an earthquake, such WiFi—enabled cockroaches could locate and determine the condition of buried victims.

WiFi--enabled cockroaches at North Carolina State University.

Researchers at MIT recently showed that using existing WiFi they could track people's movements, measure their breathing rates, measure their size and weight, identify them based on previously acquired 'WiFi templates', and detect if they had fallen down. These capabilities would seem extremely valuable in monitoring the elderly as well as many other new applications.

Finally, other researchers at MIT have demonstrated a new kind of antenna that can capture electromagnetic waves, including Wi-Fi signals, as AC waveforms. The antenna is connected to a novel device made out of a two-dimensional semiconductor just a few atoms thick. The AC signal travels into the semiconductor which converts it into a DC voltage that could be used to power low-power electronic circuits or recharge batteries.

Does WiFi pose any health risks? The World Health Organization says 'no', and a recent British study concluded that being exposed to WiFi for 1 year is roughly equivalent to the amount of radio-wave exposure in a 20-minute cell-phone conversation—which itself is believed to be totally harmless.

WiFi technology has contributed enormously to our culture and future developments promise still more benefits.

STEALTH MILITARY TECHNOLOGY

In 1943, toward the end of the 2nd World War, the U.S. Army's Air Tactical Service Command (ATSC) met with the Lockheed Aircraft Corporation to develop a jet fighter to counter the threat of the new jet fighters flown by the Nazi Lutwaffe. The top-secret research laboratory to do this was named 'Skunk Works', directed by a young aeronautical engineer named Clarence ('Kelly') Johnson. The first U.S. jet fighter, the 'XP-80 Shooting Star', was designed, built, and delivered to the ATSC in just 143 days. In 1976, Skunk Works developed the F-117, the first U.S. stealth fighter.

Stealth aircraft technology (whether applied to aircraft, naval vessels, or land vehicles) can be divided into 5 categories: 1) aircraft shape; 2) nanotechnology coating of metal surfaces; 3) reduction of engine noise; 4) reduction of engine heat output; and 5) reduction in the radar signature of carried ordinance. (1), (2), and (5) are connected with making an aircraft invisible to radar, while (3) and (4) make an aircraft more difficult to target using heat and acoustic sensing missiles. Current stealth designs are far superior to those of the 1970's-1990's, mainly as a result of the availability of much higher computing power.

Radar detection starts with a transmitter sending out radar pulses. If any radar pulses encounter a target, the angle at which they strike the target's surface is the same as the angle the radar pulse is reflected. Most radar systems consist of a radar transmitter and receiver at the same location, so only that part of the target's surface that reflects radar exactly back to the radar transmitter location can be detected. However, when a non-stealth target has many curved surfaces, it is highly likely that at least some radar reflections will reach the receiver and the target will be detected.

A basic stealth design uses many flat surfaces that make it unlikely that any one surface will be correctly oriented to reflect radar back to the location of the radar transmitter and receiver. Such designs also avoid 'retro-corners', which are neighboring surfaces angled exactly 90^0 to each other.

The diagram below illustrates a 90^0 retro-corner with two incident radar beams—one with coarse dashes and the other with fine dashes, striking the horizontal retro-corner surface at two very different angles. However, each of the two reflected radar beams emerges parallel to its corresponding incident radar beam, and hence each reflected radar beam can effectively return to the radar receiver rendering the target detectable.

In the case of a *non*-90⁰ retro-corner, the reflected radar beams are not parallel to their corresponding incident radar

Diagram of a retro-corner design.

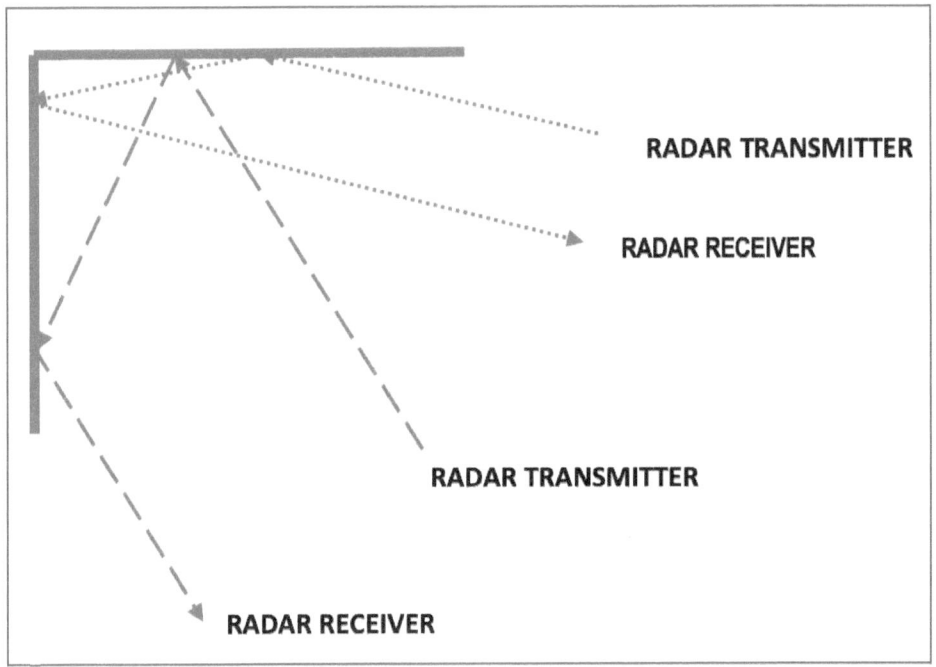

beams, and so cannot return to the radar receiver's location; and consequently the target becomes undetectable. This is illustrated in the picture below.

In modern stealth designs, aircraft surfaces, generally made of aluminum or titanium alloys, are coated with nanomaterials that strongly absorb radar. High-speed flight, however, can damage such coating materials and decrease

Diagram of a non-retro-corner design.

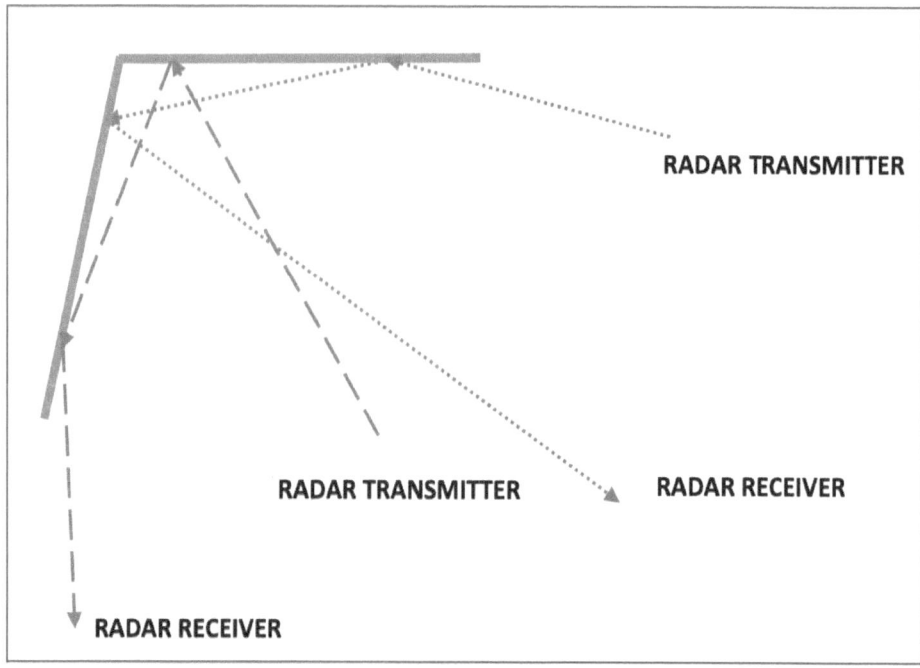

their effectiveness, resulting in nano-coated surfaces often needing to be recoated between missions.

To render stealth aircraft undetectable to missile targeting systems, engines are buried deeply within the aircraft's fuselage and their exhaust is cooled before being released outside the aircraft. Stealth aircraft engines are also designed for lower temperature and reduced sound intensity. Ordinance carried by stealth aircraft (bombs, missiles, etc.) is stored inside the aircraft's fuselage and only deployed to the outside of the fuselage immediately prior to release. During these short periods a stealth aircraft becomes vulnerable to detection.

In general, aircraft performance and stealth design are mutually incompatible. Stealth aircraft tend to be slower, less maneuverable, and less aerodynamic than their non-stealth counterparts, so they are used for specific missions where avoidance of detection is much more critical than superior flight performance.

However, the cost of stealth aircraft is extremely high— about 2 billion dollars for the stealth B-2 bomber. One reason for this is that due to its relatively inferior aerodynamic design, the stealth B-2 is more difficult to fly and therefore depends much more heavily on extremely sophisticated—and costly— electronic flight systems. The picture below shows a U.S. B-2 Stealth Bomber.

Since stealth aircraft send radar reflections away from the radar source and associated receiver, multiple synchronized radar receivers deployed away from the original radar source may detect such radar reflections which can partly defeat the stealth design. Based on this concept, the U.S. and the U.K. have developed effective stealth aircraft detection systems that use radio signals from cell-phone towers. Radar receivers deployed away from the original radar source may detect such radar reflections which can partly defeat the stealth design.

U.S. B-2 Stealth Bomber.

The three pictures below show the U.S. stealth destroyer 'Zumwald', an experimental Lockheed stealth vessel the 'Sea Shadow IX-529', and a Polish NATO stealth tank the 'PL-01'.

Stealth technology has progressed a long way since its origins in the post-2nd world war development by Skunk Works. Perhaps the next evolution in stealth technology could be total invisibility. I don't think so, but who knows what super-secret technologies may be released in the future.

U.S. Stealth Destroyer Zumwald.

U.S. Stealth vessel IX-529 Sea Shadow.

Polish NATO stealth tank PL-01.

UNDERWATER SUPERCAVITATION

'Underwater supercavitation' is a phenomenon that can potentially allow objects to travel underwater near the speed of sound in water -- about 3,350 mph.

Supercavitation occurs when a thin layer of air surrounds an object traveling underwater so that the resultingly reduced drag allows it to potentially attain speeds it could only achieve if it were actually traveling through air. At 'water-Mach-1' speed, an underwater trip from Liverpool in the U.K. to New York would take only about 1 hour. Is such travel within the realm of scientific possibility? The answer is a qualified 'yes'.

In 1960, the USSR started a project to develop a high-speed torpedo, 4-5 times faster than traditional modern torpedoes. This device was put into production in 1978. In 2004, a German weapons manufacturer announced a supercavitation torpedo called the 'Barracuda' capable of reaching speeds of 250 mph. Not to be outdone, in 2005 the U.S. Defense Advanced Research Projects Agency (DARPA) announced the Underwater Express Research Program to build a new class of submarine capable of speeds up to 115 mph-- almost 4 times faster than the fastest current submarines.

Supercavitation can occur naturally when an object moving rapidly underwater reaches a certain threshold speed--analogous to an aircraft breaking the sound barrier--or it can be artificially induced by continually injecting a gas around the surface of such an object. The speed required for natural supercavitation can in principle be achieved by rocket propulsion, but induced supercavitation can occur at far lower speeds. The Russian torpedo mentioned above uses rocket propulsion to induce natural supercavitation as well as additionally using gas-induced supercavitation.

For objects traveling at supercavitation speeds, steering clearly becomes problematic. Various ingenious solutions have been implemented: off-axis tilting of the object's nose, variable patterns of gas injection over the object's surface, and multiple retractable fins.

Technologies already exist that may enable future travelers to conquer the vastness of oceans at outlandishly higher speeds than are currently possible.

The first picture below shows the Guardian, an experimental underwater vessel developed by Juliet Marine Systems of Portsmouth, NH that uses induced supercavitation to achieve exceptionally high underwater speeds.

The next picture below shows an artist's conception of a new class of supercavitating submarine. This Chinese design, however, is not intended for military purposes but for transporting passengers. A Shanghai-to-San Francisco underwater trip taking 100 minutes was mentioned in the announcement.

The Guardian, a supercavitation submersible vessel from Juliet Marine Systems.

But even if passengers were ever transported at such high speeds, no known technological solutions have yet been proposed to prevent the loss of their luggage.

Chinese proposed design of a supercavitating submarine vessel intended for transportation of (extremely wealthy) passengers.

DEFINITION OF TIME AND ATOMIC CLOCKS

The international time standard—often referred to as the 'atomic clock'—is located at the National Institute of Science & Technology (NIST) in Fort Collins, Colorado. Much simpler technologies were used in earlier days. In 1656, Christian Huygens invented the pendulum clock. It had a pendulum that gave it a swing of one second, and was the first clock that could *reproducibly* keep time in seconds--although it did not *define* the length of a second. In 1956, for the first time, the second was defined as the fraction 1/31,556,925,9747 of the length of a tropical year at the equator, and this definition was adopted as part of the International System of Units in 1960.

In 1967, the second was redefined as the *duration of 1/9,192,631,770 of the wavelengths of the transition frequency between two hyperfine levels of the ground state of a Cesium-133 atom*. The isotope Cs-133 is the central component of the atomic clock. The Cs nucleus has 55 electrons that zip around it in elliptical orbits. These orbits also represent energy levels in which these 55 electrons are bound to the nucleus, with the highest being near the nucleus and the lowest being in the outer peripheries of the atom. The energy level closest to the nucleus is split into two so-called 'hyperfine' levels. If a microwave with *exactly* the same energy as the energy difference between these two hyperfine levels strikes the Cs atom, some electrons in the lower hyperfine level could be flipped into the higher hyperfine level. This so-called 'resonance' frequency is exactly 9,192,631,770 wavelengths/second. If we were able to count these wavelengths we would be able to exactly define the length of a second.

    The NIST folks invented an awesome device to count these wavelengths called an 'atomic fountain'. Some Cs-133 is first heated in a vacuum to produce a gas of individual Cs atoms. A series of 6 infrared lasers, their beams intersecting at a single point, nudge individual Cs atoms toward this focal point creating a small 'bunch' of Cs atoms. One of the six lasers--the one pointing upwards--then nudges the bunch of Cs atoms upwards against gravity into the opening of a microwave cavity. The bundle of atoms passes vertically through the microwave cavity and leaves through the top. Then, it immediately starts to float back down through the microwave cavity under the force of gravity (hence the name 'fountain'). After reentering the microwave cavity from the top, the bunch of Cs atoms is exposed to microwaves at roughly the Cs atom's resonance frequency. The bunch of Cs atoms then leaves the microwave cavity, this time from the bottom, and passes through the beam of a 'probe' laser that delivers infrared pulses of light to electrons located in the lower energy hyperfine levels of the Cs atoms knocking them back to the higher-energy hyperfine levels. Every time this happens, the Cs atoms emit small flashes of light called 'fluorescence' that can be counted by a detector. If the microwave frequency in the microwave cavity was exactly the same as the Cs atom's resonance frequency, the signal in the detector would be at a maximum, since the largest number of Cs atoms at the lower energy hyperfine levels would have been produced within the microwave cavity. The microwave frequency is then swept to identify the largest signal, and that frequency is the resonance frequency of the Cs atom. The microwave generator then counts the time required for the 9,192,631,770 wavelengths of the Cs resonance frequency to pass by, and that is, by modern definition, the length of time of a standard second.

    But why bother with this fountain thing? Because at the top of the atomic fountain, because the Cs atoms cease to have any motion, their temperature is reduced close to absolute zero, and that is the temperature at which the 9,192,631,770 wavelengths/second is valid. To achieve an accuracy of 1 second in 1.4 million years you have to worry about minor details like that. The picture below is a diagram of the U.S. Cs-133 atomic clock located at Fort Collins, Colorado.

The atomic clock is in reality a highly complex device and serves to provide the official scientific definition of the second. But what if a comparable level of timekeeping accuracy is required by you and me?

Using the principle of the 9,192,631,770 wavelengths/second resonance frequency of the Cs-133 atom, a solid state version of an atomic clock that could be worn on the wrist was developed by the Bathes company in 2013. A certain degree of sacrifice in accuracy was required in converting the original Fort Collins design as described above into an equivalent solid state device, but the Bathes Cs-133 wristwatch is still accurate to 1 second in 1,000 years.

In consideration of those of us who cannot afford a Bathes Cs-133 atomic watch but really need to be on time for appointments, the Citizen watch company in 1993 developed a 'poor man's' atomic watch. Such atomic watches are typically quartz controlled watches that also contain hardware and special software to receive synchronization radio signals produced by the U.S. atomic clock laboratory in Fort Collins, Colorado.

Diagram of a Cesium-133 Atomic Clock located at NIST.

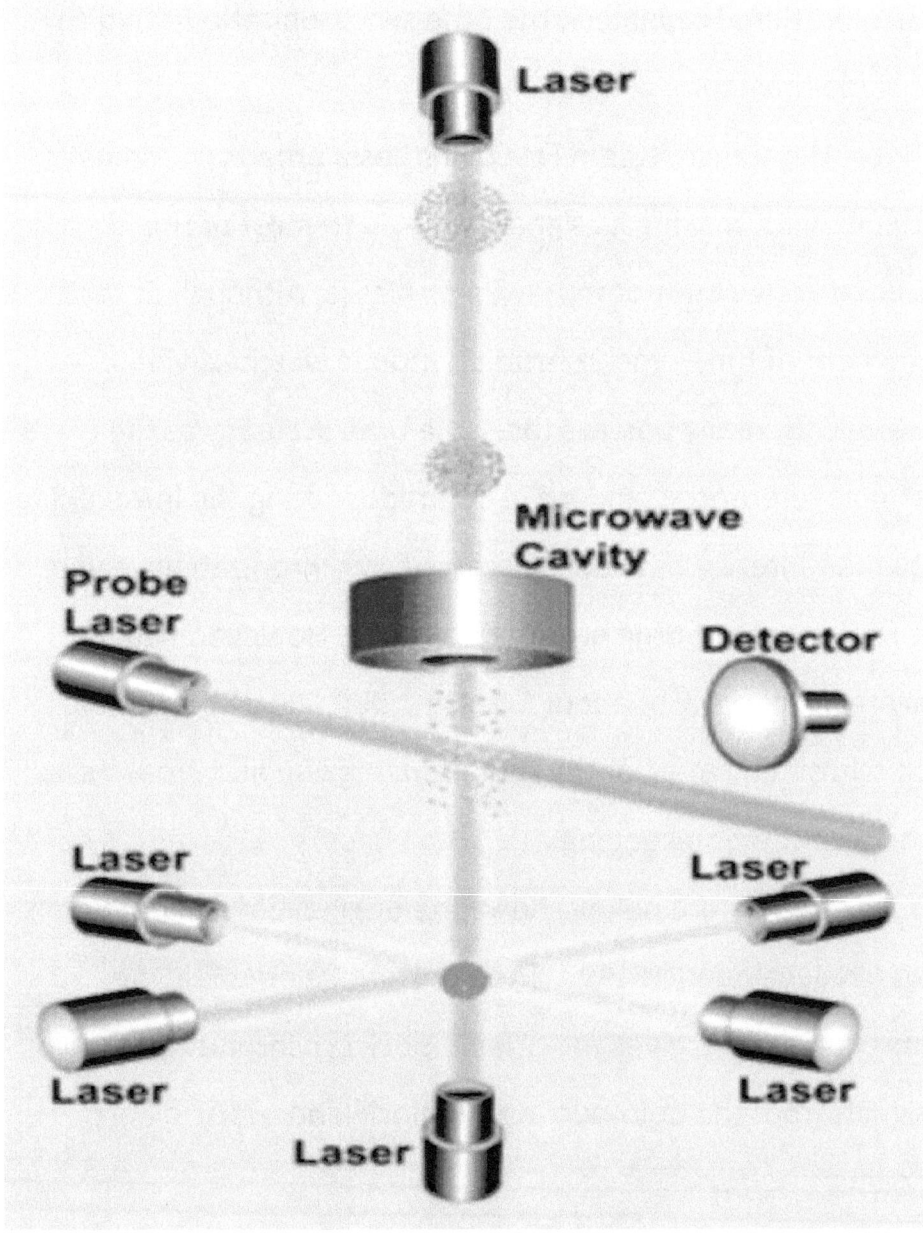

Atomic watches receive the atomic clock synchronization signals once every 24 hours. The time synchronization signal accounts for daylight saving time, leap years, and leap seconds. And if the atomic watch also contains GPS locational capability, the time synchronization can also include time-zone differences.

The transmitter in Ft. Collins has a broadcast radius of 1,864 miles, making its signal available to most of the United States but not to Hawaii or Alaska. Although at the moment of time synchronization atomic watches will transcendentally possess the same time accuracy as the Cs-133 atomic reference standard in Colorado, during the intervening 24 hour interval between successive synchronizations a very minor degree of time error can creep in. However, with an average accuracy of about 1 sec in 1,000,000 years, atomic watches fulfill most of the requirements you and I may have for accurate timekeeping.

The picture below shows the original Citizen atomic watch developed in 1993. The horizontal element is the antenna that receives the atomic clock synchronization radio signals from the Colorado atomic clock laboratory in Fort Collins.

The first atomic wristwatch developed by Citizen in 1993.

THE IRAN NUCLEAR AGREEMENT

Iran has always claimed that its program for the enrichment of uranium is a necessary step toward providing various civilian services, such as radioisotopes for nuclear medicine, a civilian nuclear energy program, and various civilian nuclear research programs. However, this has always clashed with widespread international belief that these claims have simply been a cover for a much more nefarious goal: of joining the select nuclear club of nations that possess nuclear weapons.

The Joint Comprehensive Plan of Action (JCPOA) was a nuclear non-proliferation agreement between the U.S., the U.K., France, Germany, the European Union, Russia, China, and Iran, as well as a 'sub-agreement' between Iran and the International Atomic Energy Agency (IAEA) for the narrower purpose of inspection and verification. The JCPOA had originally been negotiated by the Obama administration and was signed by all the above on July 14th, 2015.

One of Donald Trump's campaign promises was that he would withdraw from the JCPOA because it was a 'completely one-sided agreement—overwhelmingly benefitting Iran'. President Obama's retort to this criticism was that it may not have been the most favorable agreement for the U.S., but it was much better than nothing. In return for abiding by the JCPOA agreement, Iran would receive relief from U.S., European Union, and United Nations Security Council economic sanctions, which were severely crippling Iran's economy, despite the fact that a number of nations, primarily Russia, were overtly bypassing these sanctions.

On October 13th, 2017, Donald Trump announced that the United States would not provide the 'annual certification' for the JCPOA, a procedure required under U.S. law, although he stopped short of terminating the agreement altogether. However, On May 8th, 2018, Donald Trump announced the U.S.'s total withdrawal from the JCPOA agreement, leaving the remaining nation signatories in somewhat of a mess.

The following are the primary commitments required of Iran under the JCPOA agreement:

1) *Reduce the stockpile of medium-enriched uranium down to 300 kg*

Uranium is an element consisting of a number of different isotopes. Isotopes are nuclear variants of an element that despite having identical *chemical* properties, possess different *nuclear* properties. Two important isotopes of the element uranium for this discussion are uranium-235 and uranium-238. 'Natural' uranium, as it is mined from the ground, is only 0.7% uranium-235 and about 99% uranium-238. Although both these uranium isotopes are moderately radioactive, U-235 is the only 'fissile' isotope found in nature. A fissile isotope is one whose nucleus can be made to break apart--or 'fission'--following the absorption of a nuclear particle called a neutron. When a U-235 nucleus fissions, it re-emits on average 2-3 neutrons and a tremendous amount of energy, mainly as heat, light, and ionizing radiation (the potentially dangerous kind), as well as a large number of potentially dangerous byproduct isotopes. Each of the 2-3 *outgoing* neutrons from the first fission event can now be considered as 2-3 new *incoming* neutrons that can initiate 2-3 more U-253 fission events. Therefore, at each step in this process, the amount of energy released doubles or triples until all the U-235 has been 'burned' up.

Boron-10, like U-235, also hungrily absorbs neutrons, but unlike U-235, it does not re-emit any. Therefore, if B-10 is mixed with U-235, it competes with it in absorbing neutrons so there are fewer neutrons available to 'feed' the U-235 fission reactions. With just the right amount of B-10, the U-235 fission-rate (in fissions/sec) remains constant with time and a constant controlled production of fission energy results. The B-10 is incorporated into a series of 'control-rods' that can be inserted into channels in the U-235 fuel by varying amounts. Full insertion absorbs more neutrons than is necessary to sustain the U-235 fission-rate and is used for reactor shutdown. Partial insertion can be adjusted to produce a constant U-235 fission-rate for safe, constant energy production. No insertion may result in a progressive increase in fission-rate, which if not interrupted can result in a serious nuclear accident such as a core meltdown.

However, to make a nuclear weapon, we want the maximum amount of fission energy to be produced in the shortest possible time—i.e., we need an 'explosion', which means we need to let the fission rate increase unchecked as rapidly as possible until the entire U-235 has been burned up.

I mentioned earlier that about 99% of the uranium mined is the isotope U-238. U-238, which is not fissile, also absorbs neutrons, but much less avidly than U-235. But because there is so much of the U-238 in natural mined uranium, it acts like a deeply inserted boron-10 control rod; i.e., it can stop, or at least greatly slow down, the U-235 fission process. So 'raw' mined uranium cannot easily be used either for fueling a nuclear reactor or for making a nuclear weapon. What is usually done is to replace some of the U-238 with U-235 to enable the fission process to progress at a constant rate with time--as in a nuclear reactor, or to permit it to grow rapidly with time unchecked--as in a nuclear weapon. Replacing U-238 with U-235 is called uranium 'enrichment'.

For use in nuclear reactors, 'low-enrichment' U-235 enriched by 0.9-2% is desirable (compared to the mined level enrichment of 0.7%) and is often referred to as 'medium enrichment'. For nuclear weapons, U-235 enrichment must be increased to at least 20%, but more realistically to 90% or more. When the enrichment level of uranium is higher than 20%, uranium is referred to as 'highly-enriched' or 'weapons grade'—and is what the JCPOA was intended to limit Iran from producing. Limiting Iran's stockpile of medium-enriched uranium to 300 kg would require more time for Iran to enrich enough uranium to weapons-grade to manufacture one or more nuclear weapons.

2) *Cut by 98% its stockpile of low-enriched uranium*

Although there would seem to be no immediate risk for Iran to possess low-enriched uranium, the problem is that with the availability of appropriate technology, low-enriched uranium can be converted to highly-enriched, or weapons-grade uranium. Therefore, requiring a 98% reduction of Iran's stockpile of low-enriched uranium would remove much of the uranium that otherwise could be enriched to weapons-grade.

3) *Reduce by two-thirds the number of its gas centrifuges*

Gas centrifuges are the most common technology used to enrich uranium. The concept is quite simple, as illustrated in the picture below, but the technology is highly sophisticated.

Diagram of U-235 enrichment centrifuge.

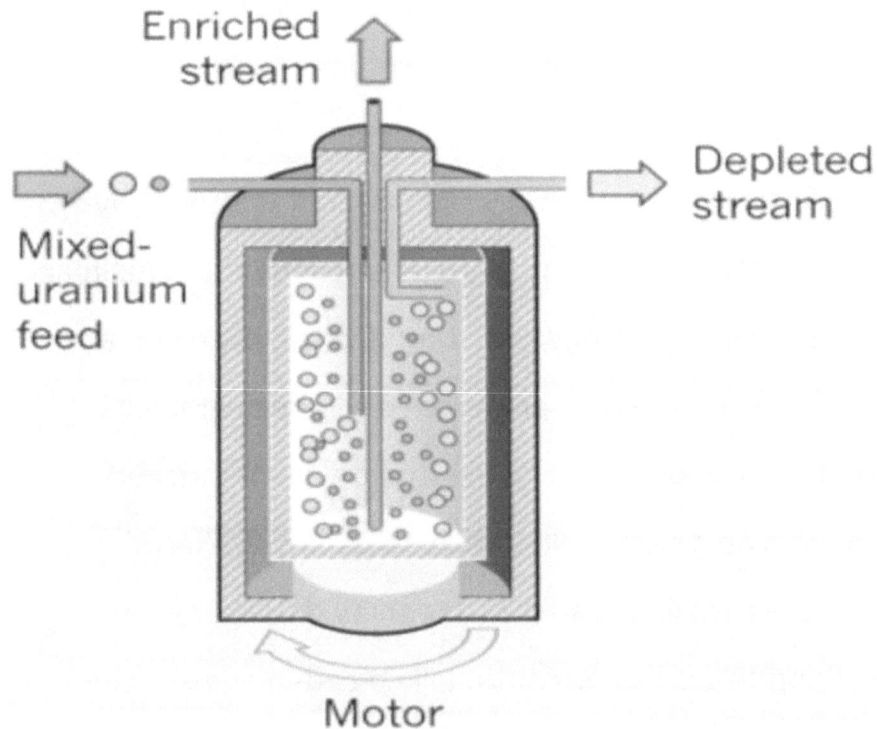

Unenriched uranium is mixed with fluorine and converted to gaseous form as uranium-hexafluoride gas, which is injected into a long tube that spins very fast around its long axis. Because atoms of U-238 are very slightly heavier than those of U-235—just 1.3% heavier—this is enough for them to be spun more quickly out toward the periphery of the centrifuge tube, so that gas atoms of U-238 and U-235 will become progressively separated from each other, and the uranium gas extracted from the center of the gas centrifuge will be more highly enriched in U-235. The longer the uranium remains in the centrifuge, the more enriched the extracted uranium becomes. The larger the volume of the centrifuge or the larger the total number of centrifuges operating concurrently, the faster would be the enrichment process.

Therefore, the JCPOA's requirement to reduce by 2/3 the number of Iran's centrifuges, and to limit the remaining ones to older technology, was intended to slow down any attempts by Iran to further enrich to weapons-grade its permitted residual stores of low and medium-enriched uranium. The picture below shows a centrifuge 'farm' in Iran. Centrifuges are linked so that unenriched uranium hexafluoride gas can be continually piped-in and enriched U-

235 hexafluoride gas continually extracted from all centrifuges automatically. See person on left side of picture for scale.

Gas centrifuge 'farm' in Iran.

4) Only enrich uranium up to a level of 3.67%

Uranium enriched to a level of 3.67% is a 'safe' enrichment level to be used as reactor fuel, but far short of the 20% or higher enrichment level required for nuclear weapons. Limiting the permitted uranium enrichment to 3.67% was another approach to prevent Iran enriching its residual uranium stockpile to weapons-grade.

5) Not build any new heavy-water facilities

U-235 is not the only fissile isotope—although the only one that occurs in nature. Nuclear weapons can also be manufactured—in fact more efficiently—using another fissile isotope, plutonium-239; but Pu-239 does not exist in nature and first needs to be 'made' from naturally occurring U-238. To make Pu-239 requires U-238 to be bombarded by large numbers of neutrons—as it normally would be within a nuclear reactor's fuel. The concentration of Pu-239 increases in the reactor fuel with time, and can then be separated from the byproduct isotopes in burnt-up reactor fuel. Nuclear reactors that can produce large enough quantities of neutrons for this to be practical preferentially use heavy-water rather than light-water to submerge the uranium fuel rods. Therefore, restricting Iran's capabilities to manufacture heavy-water would slow down its ability to make Pu-239.

6) Limit uranium enrichment activities to a single facility using only first- generation centrifuges

The requirement was again intended to greatly slow-down Iran's ability to produce weapons-grade uranium. Centrifuges, despite their fairly simple operating concepts, are in fact difficult to manufacture efficiently due to the high required rotation speed and the necessary reliability. First-generation centrifuges are far less efficient in enriching uranium than modern centrifuges, which is why there was the necessity to limit Iran's access to modern centrifuges.

7) *All other enrichment facilities would be converted to avoid further proliferation risk*

This catch-all condition would theoretically preclude Iran from operating any additional enrichment facilities that fell under the IAEA's radar.

If Iran enriches its uranium to less than 20%, it cannot utilize it for weapons. However, if that is not the case, an inescapable conclusion must be that it intends to use the weapons-grade uranium for nuclear weapons manufacture.

Many years prior to the JCPOA agreement, a deal brokered by the U.N. and the U.S. to conditionally relieve Iran of the then existing sanctions was to permit Iran's 'civilian' nuclear program to continue, but to prevent the enrichment of its uranium to more than 20%, which again would have theoretically prevented it from producing weapons-grade uranium. Under this plan, Iran was required to export its low and medium-enriched uranium assets to Russia, where it would be incorporated directly into reactor fuel rods, and then re-exported back to Iran. Once the uranium—-well over a ton of it—-had been converted into reactor fuel rods, it would be impossible to enrich it further to weapons-grade level. Unfortunately, this very clever plan was rejected by Iran—and considering current politics it would have been questionable whether Russia would have adhered to their required role under this agreement.

With Donald Trump's withdrawal from the JCPOA agreement, the remaining participants in the JCPOA have been presented with a complicated dilemma. Only the U.S., Russia, and China possess the military resources to react to a contravention of the JCPOA agreement by Iran; and it is questionable what actions Russia or China would take in such an eventuality. With Donald Trump's withdrawal from the JCPOA and, more recently, his withdrawal of the U.S. military presence in Syria, Russia has been given a green light to revert to its political goal of destabilizing the delicate middle-eastern balance of power.

However, President-Elect Joseph Biden has committed the U.S. to rejoin the JCPOA agreement as one of his first presidential directives in January 2021, and in doing so will have averted one of the most dangerous military situations the U.S. has faced since the Cuban missile crisis.

RED MERCURY HOAX

Some years ago, C.J. Chivers wrote a fascinating article in the New York Times about the interest that ISIS had in obtaining supplies of a so-called 'doomsday material of dreams' called *red mercury*. For a terrorist organization to purchase a suitcase-sized nuclear weapon on the black market is next to impossible or at least extraordinarily expensive. However, something similar--almost as deadly, readily obtainable, and far less expensive, would be warmly welcomed by terrorists worldwide.

Shortly after the demise of the Soviet Union in 1990, there were some unsourced reports that the collapsing Soviet government decided to divest itself of all supplies of a material called *red mercury* to protect the emerging Russian Federation from terrorist elements that might use this so-called doomsday material to manufacture small suitcase-sized weapons that would match the explosive power of a small nuclear bomb. The story goes that red mercury was eliminated from the Russian Federation's military storage sites by packaging small quantities of it in Singer Sewing machines, which were then exported out of the Russian Federation onto the international terrorist arms market. Terrorist organizations, primarily ISIS, then purchased every Singer sewing machine they could lay their hands on to extract the Red Mercury supposedly contained inside.

Now even if this story so far sounds utterly ludicrous, read on. If red mercury was in fact the doomsday material that its reputation claimed, acquiring it would alter the capabilities of terrorist organizations throughout the world. In fact, red mercury-containing sewing machines are still being sold on the international terrorism market for approximately $1,800,000 per kilogram of red mercury.

Red mercury is simply formed of the elements mercury and iodine, compounded as *mercuric-iodide.* Does mercuric-iodide possess any explosive capability? Definitely not. No isotopes of either mercury or iodine are 'fissile', which means that they cannot be split by absorbing neutrons as uranium-235 and plutonium-239 do to produce a nuclear explosion. It is the splitting of a heavy fissile isotope into two lighter isotopes that releases the tremendous yield of energy that is the basis of the fission explosion. Nor can Red Mercury have any role in nuclear fusion, since nuclear fusion requires the coalescence, only possible under extremely high temperatures and pressures, of two very *light* isotopes, such as deuterium, tritium, or lithium which are about 50-100 times lighter than mercury or iodine.

Let's segue to a quick explanation of how fusion bombs work—usually referred to as 'thermonuclear' or 'hydrogen' bombs. In the cold war days, when the U.S. believed that there was a high likelihood that its H-bombs and nuclear warheads might need to be used, the light isotopes that were used as fusion fuel were deuterium (a hydrogen atom containing an additional neutron) and tritium (a hydrogen atom containing twp additional neutrons). H-bombs use a primary fission stage, consisting of a uranium-235 or plutonium-239 fission explosion, to create the extremely high temperatures and pressures required for the secondary stage of the weapon–the fusion stage–to be set off. However, tritium is a radioisotope with a half-life of 12.5 years, so every 12.5 years, the explosive power of the U.S. and Soviet thermonuclear weapons arsenals was diminished by 50% and had to be recharged with tritium. Recharging the weapons with fresh tritium was not a simple procedure because the thousands of H-bombs and nuclear warheads were geographically spread out over the U.S. and the Soviet Union. Also, the slow production-rate of fresh tritium in the limited number of military nuclear reactors hindered the process.

Towards the end of the cold war, both the U.S. and the Soviet Union developed an elegant solution to the tritium decay problem. Instead of using deuterium and tritium as the fusion isotopes, the isotopes lithium-6 and deuterium, in the chemical form of *lithium-deuteride*, were used instead. Lithium-6 strongly absorbs neutrons generated by the primary fission stage of an H-bomb explosion, and in so doing produces tritium. So the supply of tritium for the fusion stage of the weapons was made immediately available at the moment the weapons were triggered, obviating the need for periodic in-the-field recharging with fresh tritium.

The story went that red mercury, when mixed with conventional explosive material, could replace the fission stage of a thermonuclear weapon much more simply and cheaply than an actual U-235 or Pu-239 fission device.

There is credible evidence that within the Soviet Union's defense establishment, red mercury was the code name used by physicists for—you guessed it--lithium-deuteride. But why would the Soviets produce such a monumental hoax? They didn't. The origin of red mercury being a potential weapon of mass destruction was initially floated by unknown sources, but

Samuel T. Cohen, a retired American physicist who worked on building the first atomic bomb at Los Alamos, provided scientific support to the red mercury hoax. Why he did this is still somewhat of a mystery, but it does explain why the belief in red mercury's claimed properties was so strong among terrorist organizations. It is also not clear who planted mercuric-iodide inside the hundreds of sewing machines, but this hoax netted many middle-eastern arms brokers huge windfalls.

The only positive fallout (no pun intended) of this story is that ISIS, the primary customers for Singer sewing machines containing red mercury, have squandered enormous amounts of their financial resources that would otherwise have bankrolled other terrorist activities.

A marketplace for red mercury-containing sewing machines.

HORSESHOES AND ABRAMS TANKS

Horses are heavy animals, weighing around 1,000-2,000 lb, depending on their breed, compared to about 140,000 lb for an Abrams tank. Frequently, studs are attached to a horse's shoes for the purpose of improving traction during winter. However, such studs are potentially subject to transient compressive stresses that can be as high as 2,000 lb/in^2. Such mechanical stresses are more than steel or aluminum—the most common materials that horseshoes are made of—can tolerate without excessive wear. Therefore, horses are frequently outfitted during the winter season with 'Borium studs' attached to their shoes so they can safely navigate icy pastures and roads. Contrary to prevailing belief, Borium, which is considered to be hard enough to tolerate the high mechanical stresses mentioned above, is neither an element nor a metal but a brand name for a very tough metal alloy called *tungsten-carbide*. Tungsten-carbide is one of the hardest materials known that can only be scratched by diamonds.

But there are other materials that have specific properties related to the concept of 'hardness' or 'density' that I would like to mention. The weirdest of these is probably *Bohrium* (spelled with an 'h'), an element artificially created in a laboratory in 1976 by Russian scientists and named in honor of the Danish physicist Niels Bohr. Unfortunately, all isotopes of Bohrium are highly radioactive and would normally be considered extremely hazardous were it not for the fact that only about 10 atoms of Bohrium have ever been created at any one time! But *if* a solid version of bohrium could be produced, it would be an extraordinarily heavy metal with a physical density of about 37 g/cm^3, making it the third densest of the 118 known elements. In comparison, tungsten and uranium have densities of only about 19 g/cm^3, while lead has a density of 11 g/cm^3—a veritable 'Jello' among dense metals.

Incidentally, no one has yet measured the density of *Kryptonite*, but in the fictional world of *Superman* it is expected to be somewhere close to infinity.

But physical density does not necessarily predict the hardness of a material. For example, gold, like tungsten and uranium, has a density of 19 g/cm^3, and yet can be scratched with a fingernail; while diamond, with a density of only 3-4 g/cm^3, is employed (beyond its romantic uses) to cut, grind, and inscribe very hard materials such as tungsten-carbide.

Dense or hard materials are useful for a number of reasons: hard materials are resistant to wear while dense materials can provide weight within a small physical space, like ballast used to stabilize aircraft, submarines, and other ocean-going vessels, or used as radiation shielding in military applications or in radiation therapy departments in hospitals.

Many years ago, the U.S. military developed a special interest in a very hard material called 'depleted uranium' (or DU). 'Depleted' refers to the virtually complete removal of the 0.7% of radioactive uranium-235 present in the originally mined uranium ore. DU consists of over 99.9% uranium-238, a very minimally radioactive isotope of uranium. In addition to being very dense and hard, DU also has some unique properties that have important military value, including an almost complete lack of 'flow' if installed without external support, a low penetrability-to-weight ratio when deployed to protect personnel against incoming shells or missiles, and exceptional X- and gamma-ray shielding properties. These properties make DU a desirable choice for fabricating defensive armor plate for Abrams tanks and other land and sea-going military vehicles, providing protection against penetration of incoming shells and missiles, as well as a high level of protection against certain types of radiation threat while minimizing weight.

More controversially, DU is also used in armor-piercing projectiles, principally as anti-tank and anti-ship weapons. For that purpose it has two further unique properties: it is mechanically 'self-sharpening' and chemically 'self-incendiary' when exposed to high mechanical impact pressures. This helps a DU-enabled shell or missile to penetrate thick armor plate more easily and then maximize the resulting damage produced inside the target.

Some time ago a number of concerns surfaced about the use of DU by U.S. and NATO forces and its potential long-term health risks to civilian and military populations. Kidney, brain, liver, heart, and some other organs were shown to be damaged by internal exposure to DU aerosols, since in addition to being minimally radioactive DU is extremely chemically toxic. Airborne DU dust produced by the impact and combustion of DU munitions can contaminate wide areas around a target impact site leading to the potential inhalation of DU dust by civilians and military personnel. In 2003, during a three-week period of the Iraq war, 1,000-2,000 metric tons of DU munitions were used by U.S. forces, mostly in urban environments. NATO forces have also used DU munitions in the Balkans and Kosovo.

However, the actual toxicity of DU remains a point of controversy. Studies using cultured cells and laboratory rodents have demonstrated increased rates of leukemia as well as genetic, reproductive, and neurological disorders from chronic DU exposure. In complete contrast, however, the World Health Organization has stated that "No consistent risk of reproductive, developmental, or carcinogenic effects have been reported in humans [following exposure to DU]". Clearly, the conclusion of whether DU munitions are hazardous needs to be depoliticized before it can be believed.

Even if a replacement could be found for DU in military applications, health concerns would not necessarily be alleviated. A very limited number of replacement materials for DU exist, and are essentially limited to tungsten-cobalt or tungsten-nickel-cobalt alloys. These alloys are themselves extremely carcinogenic and toxic--in fact, far more so than what is claimed for DU. So there appears to be little hope for the curtailment of DU use by the military in the foreseeable future.

The European Parliament has repeatedly passed resolutions requesting an immediate moratorium on further use of DU munitions by NATO forces, but France and Britain have consistently rejected calls for such a moratorium, maintaining that the putative health risks of DU in humans are "still unsubstantiated". The U.S. insists that DU remains a critical component of its military technology despite its potential risk to civilian populations, so it is unlikely that the use of DU munitions will be phased out anytime soon.

I hope that I have at least clarified the scientific connection between a horse and an Abrams tank.

The first picture below show a U.S. Abrams tank, which uses DU enabled ordinance and also uses DU as shielding against incoming missiles and shells. The second picture below shows the underside of a horse's shoe displaying 'Borium' (tungsten-carbide) studs to provide traction in winter with minimal wear.

U.S. Army M-1 Abrams Tank.

'Borium' (tungsten-carbide) studded horseshoes as used during winter months by the NYPD and many other horse owners.

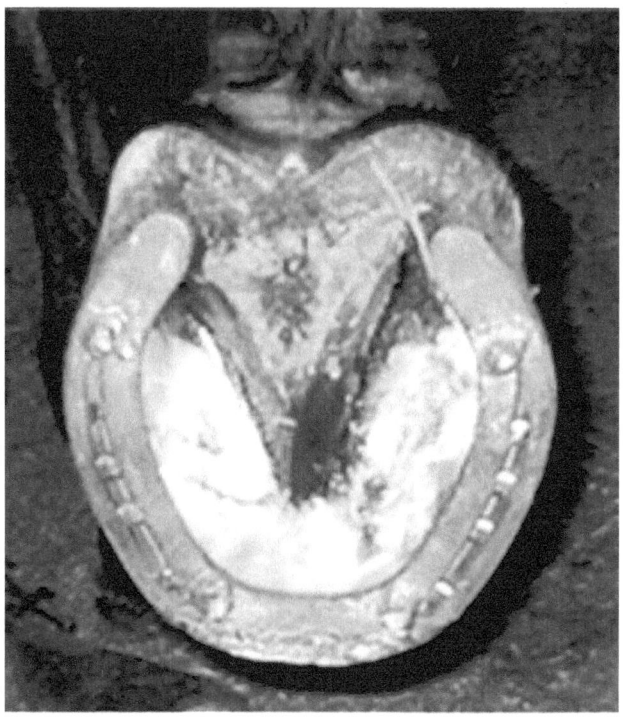

DNA FINGERPRINTING

About a 15 years ago, DNA fingerprinting (aka, DNA profiling or DNA genotyping) was just starting to be regarded as an admissible form of physical evidence in the U.S. court systems. In prior years, a lack of proper quality assurance standards and laboratory certifications for forensic DNA analysis led to numerous court challenges. Today, DNA fingerprinting has been overwhelmingly accepted as a tool to help both convict crime suspects and to exonerate many victims of erroneous convictions. Although only utilized by the country's major police departments in earlier times, today even some of the smallest police departments are able to use DNA fingerprinting as a part of their forensic investigations.

DNA fingerprinting was invented in 1984 by Alec Jeffreys, a British molecular biologist. The first DNA-supported murder conviction in the U.S. took place in 1987. DNA fingerprinting is used to link forensic evidence (such as a smear of blood) to a suspect in a criminal investigation. Perhaps even more importantly, it can be used to exonerate victims of wrongful guilty verdicts. In fact, since 1989, 25% of violent crime suspects in the U.S. had their convictions overturned as a result of DNA evidence.

DNA fingerprinting is also used to establish paternity, by anthropologists to study genetic relationships between human ethnic groups, and to predict an individual's predisposition for familialy acquired diseases such as breast cancer, multiple sclerosis, etc.

More unusual applications have included matching goatskin fragments of the Dead Sea Scrolls, verifying the remains of Josef Mengele (the Nazi 'Angel of Death'), tracing contaminated meats to their sources, debunking the claimed origin of the Shroud of Turin, and identifying American military casualties. More recently, commercial DNA analysis services such as *23-and-me* have made it possible for anyone to find out about their ethnic background, health, and ancestry at very modest cost.

The structure of DNA (**d**eoxyribo**n**ucleic-**a**cid) was discovered in 1953 by James Watson and Francis Crick. DNA is a chemical molecule within our cells that contains the so-called 'genetic code'. DNA molecules within every cell of an individual's body are identical, and it is the additional chemical code sequences strung out along the DNA molecule that determine the individual characteristics that make each person genetically unique. Two exceptions to this are human identical twins and the four offspring that armadillos always produce that each always contains identical DNA.

The shape of the DNA molecule is referred to as a 'double-helix', and a useful analogy is to compare it to a twisted step-ladder. The sides of the step-ladder are the strands of the DNA that do not contain any genetic information, while the rungs contain the chemicals that constitute the genetic code.

The rungs of the DNA ladder consist of four 'bases'. Bases are small molecules named *adenine, thymine, guanine, and cytosine* and are the building-blocks of the genetic code. About 3% of the base sequences along a DNA molecule are known as 'genes', and in human DNA there are currently about 20,000 – 25,000 identified genes. A diagram of a DNA molecule is shown below. The sequences of bases along the DNA molecule define the biological codes for growing various tissues, organs, and functions of a living organism as well as providing a unique DNA fingerprint.

To obtain a DNA fingerprint, the base sequences along the DNA are identified. Comparing the base sequences obtained from a forensic DNA sample (such as a hair root on a comb) with that of a murder suspect's DNA

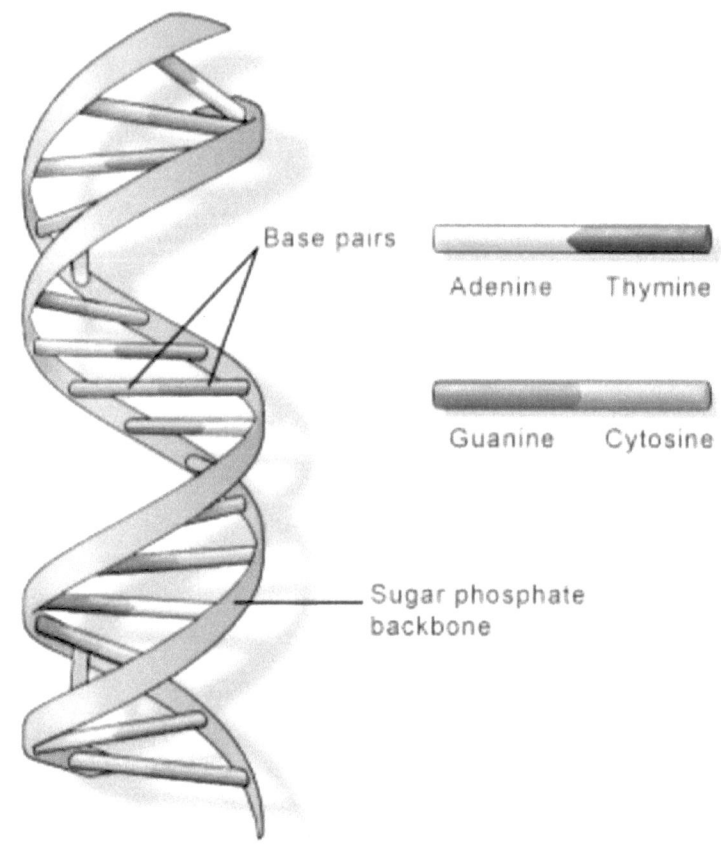

produces two DNA fingerprints that can verify or exclude a DNA match. The error rate of DNA fingerprinting depends on many factors, but is generally believed to be better than 0.001%, and much better when the DNA samples compared are genetically more distant.

Today, DNA sequencing does not require a large hugely expensive laboratory. A British company, *Oxford Nanopore*, is marketing a hand-held DNA fingerprinting device, the 'Minion', initially developed to measure the DNA fingerprints of samples on the International Space Station. A picture of the Minion device is shown below.

The MinION, a hand-held DNA fingerprinting device marketed by Oxford Nanopore Technologies. The device connects to a laptop computer.

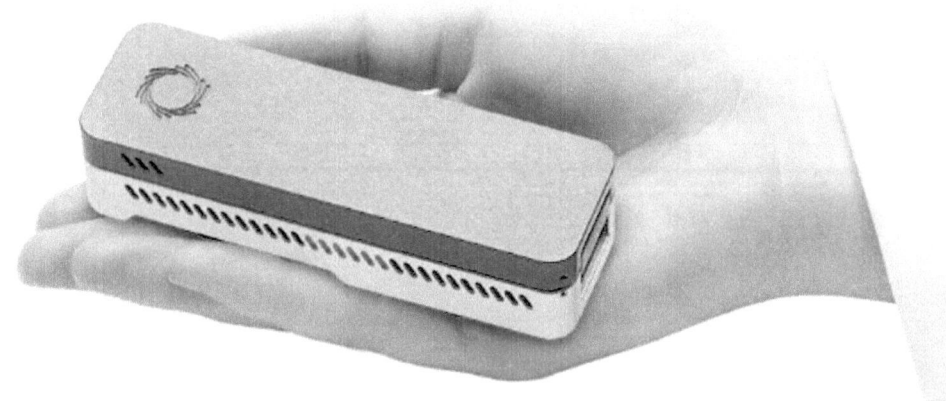

DNA fingerprinting technology is now completely accessible to even the smallest police departments. The author would like to thank the local police department in Woodstock, VT for collaborating on this article and providing him with useful background material on the forensic applications of DNA fingerprinting.

DARK MATTER

Physicists have suspected for many years—and now have lots of data to prove it--that much of what we visually perceive as empty space is in fact filled by a material that has been coined *'dark matter'*.

The mass of stars and galaxies is usually measured by observing the gravitational influence that stars or galaxies have on each other's rotational motions. These measurements can be compared to the masses derived by measuring the radiation produced by these stars and galaxies, which consists mainly of light, X-rays, and microwaves. This radiation signature is called 'luminescence'. However, masses of stars and galaxies calculated from luminescence measurements do not agree with their masses calculated from their observed rotational motions. To explain the observed rotational motions requires the presence of about 27% more mass present in space than astronomers are aware of.

This missing mass is often referred to a *'dark matter'*. The reason that dark matter is difficult to measure is that it does not interact with the universe through any means other than through its gravitational field, and such interactions are the most difficult to measure in astrophysics.

So what is dark matter? It is postulated to be composed of very heavy subatomic particles called *neutralinos,* that interact with the remainder of the universe only through their gravitational fields, and that is why they can only be detected with great difficulty by conventional experimental methods.

Is the presence of dark matter an important issue in astrophysics? Sure it is. One thing that is ascribed to the presence of dark matter is that contrary to earlier astrophysical theories, which predicted that following the primordial 'big-bang' the galaxies in the universe were continually expanding and accelerating outward, modern astrophysics has shown that in fact the reverse is the case; i.e., the universe is now gradually ceasing to expand, and eventually will execute an expansion reversal.

Should we worry that due to dark matter and dark energy there will soon be no universe remaining? No. If the big-bang happened about 14 billion years ago and the presence of dark matter eventually causes the expansion of the universe to reverse itself, the end-of-the universe should occur about 14 billion years from now. Just order another cup of latte and relax--you won't be around to witness the whimper of the end of the world (with apologies to T.S. Eliot)!

COLD FUSION: A SOLUTION TO THE WORLD'S ENERGY NEEDS?

Back in 1989, Drs. Stanley Pons and Martin Fleischmann, world-class electrochemists at the University of Utah, shook the scientific world with a press announcement that they had achieved 'cold fusion' in their laboratory. Cold fusion was coined to describe nuclear fusion occurring at room temperature, in contrast to more classic fusion that requires temperatures of hundreds of millions of degrees Celsius.

In nuclear fusion, two very light nuclei, such as deuterium (sometimes known as heavy- hydrogen), each consisting of one neutron and one proton, are forced together to create a single nucleus of a heavier isotope. In this case, the heavier isotope may be helium-3 or tritium-3 (a radioactive form of heavy-hydrogen that is even heavier than deuterium). In the first case, fusion is accompanied by the emission of a neutron; in the second case, by the emission of a proton. In contrast to nuclear *fission*, where very heavy isotopes, such as uranium-235 or plutonium-239, absorb neutrons and split into two parts releasing a tremendous amount of energy, nuclear fusion releases an even larger amount of energy—very roughly 10 times more than fission. High-temperature—or classical nuclear fusion--can occur in a thermonuclear weapon, or in a fusion reactor to generate energy for peaceful purposes. Functioning fusion reactors, however, still have eluded scientists, which is why cold fusion would be such an earth-shattering technological development.

The experimental setup for Pons & Fleischman's cold fusion experiment is remarkably simple. A water-bath is filled with heavy water (where the hydrogen is mainly deuterium). Two electrodes are submerged in the heavy water, one made of platinum the other of palladium. The two electrodes are connected to a battery. In addition, the water-bath contains a temperature sensor to measure any rise in temperature of the heavy water. A rise in temperature of the heavy water is one of the three 'markers' required to prove that fusion has occurred. A gamma spectrometer and a sensitive radiation detection system are also configured to detect the production of neutrons and the presence of radioactive tritium produced in the heavy water—the other two markers required to prove that fusion has occurred. Palladium metal sucks deuterium atoms from the heavy-water bath and chemically packs them much more closely together. Deuterium nuclei that are packed close together are more likely to fuse. But, as we know, like charges repel, so the positively-charged deuterium nuclei would strongly resist being forced together. This is where a phenomenon well known in nuclear physics called 'tunneling' has been invoked to explain cold fusion. There is in fact a very tiny statistical probability that despite trying to repel each other, very rarely two deuterium nuclei would successfully combine. Tunneling is one of the scientific explanations voiced by supporters of Pons & Fleischman to explain cold fusion. Skeptics, however, point out that the three conditions necessary to prove cold fusion did *not* occur in this experiment. Pons and Fleischman reported a rise in the temperature of the heavy water and they claim they had measured the presence of neutrons, but they did not try to measure the presence of tritium. Moreover, their data purportedly showing neutron production was believed by skeptics to be an experimental artifact. I saw those data and totally agree.

In the past 30 years, many scientists worldwide have attempted to replicate Pons and Fleischman's cold fusion experiment. Most of these research efforts, often very well-funded, have occurred in Japan. However, in no case were the three necessary conditions for proving cold fusion found to exist. Some research into cold fusion still persists, carried out by some eminent scientists, but as more time passes it is becoming less and less likely that cold fusion will prove to be a reality.

Nevertheless, *if* cold-fusion were proved to exist, the energy needs of our entire planet for the next 100 years could be met using the top 6 inches of water in Lake Superior as cold-fusion 'fuel', even given the fact that normal water contains only 0.0156% deuterium atoms. I do not reject the concept of cold fusion totally, but only point out that in the 11 years that have elapsed since its conception, no substantive progress has been reported.

Stanley Pons and Martin Fleischmann (1989).

Pons and Fleischmann's cold fusion experiment; yes, it was that simple!

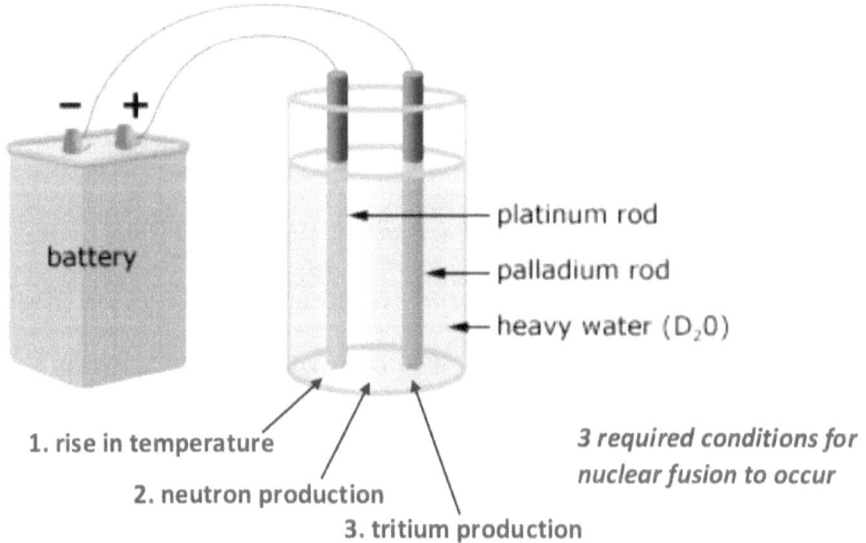

3 required conditions for nuclear fusion to occur:
1. rise in temperature
2. neutron production
3. tritium production

www.ingramcontent.com/pod-product-compliance
Lightning Source LLC
Chambersburg PA
CBHW020438220526
45464CB00002B/753